Tidal Sedimentation of the Sunderban's Thakuran Basin

Gautam Kumar Das

Tidal Sedimentation of the Sunderban's Thakuran Basin

Springer

Gautam Kumar Das
Department of Chemical Engineering
Jadavpur University
Kolkata
India

ISBN 978-3-319-44190-0 ISBN 978-3-319-44191-7 (eBook)
DOI 10.1007/978-3-319-44191-7

Library of Congress Control Number: 2016954696

Printed on acid-free paper

This Springer imprint is published by Springer Nature
The registered company is Springer International Publishing AG
The registered company address is: Gewerbestrasse 11, 6330 Cham, Switzerland

Only in the deepest silence of night the stars
smile and whisper among themselves—'vain
is this seeking! Unbroken perfection
is over all!'

Rabindranath Tagore (1912)
Gitanjali (Hymn No. LXXVIII)

To the Islanders of the Thakuran Basin

Preface

The Thakuran River is one of the many highly seasonal and tidal rivers flowing along the south coast of West Bengal, eastern India. The river is tidal for its entire 80 km length from north to south. The channel is sinuous to meandering having connections with many salt water courses. In the strict sense, the Thakuran River, like many other so-called rivers, viz. the Saptamukhi and the Matla, is a large tidal creek in this low-lying coastal plain. It has no perennial freshwater source. The river is the result of tidal incursion and retreat. It drains through the recent alluvial sediments of the Bengal Basin and flows more or less parallel to the Hugli River in the west and Matla in the east. The Thakuran has a funnel-shaped estuary at its mouth having a width of approximately 8 km. During the dry season, the river regime is controlled by tidal water, whereas during rains, the headwater discharge has a strong influence over the river regime. Much of the estuarine plain is inundated by floodwater during wet seasons, but during dry seasons, the estuarine plains are completely dried up and even register evidence of desiccation with salt encrustation. During dry times, the tidal flows accumulate in the river channels. Discontinuous stretches of mangrove forests fringe the river.

The Thakuran River is truly a major tidal creek in the low-lying, tropical coastal plains of the Ganges–Brahmaputra delta. It is fed by to-and-fro moving flood and ebb flows without any headwater supply. Geomorphologically defined areas such as mid-channel bars or flood-tidal delta, the river mouth bar or ebb-tidal delta, point bars, swash platforms, wash-over flats, and riverbanks have been identified. These areas are delineated based on studies of their physical sedimentary structures, bioturbation structures, and granulometric properties.

The Thakuran is totally a tide-influenced creek in which the process of sediment movement is accomplished by bidirectional tidal flows and wind-induced waves. The luxuriant mangrove forests here belong to a tide-dominated setting. The macrotidal mangrove-fringed creeks and estuaries are typically funnel-shaped, and the wide-opening of the Thakuran River at the seaface is no exception to this rule. The width of this water course, on the other hand, decreases exponentially with increasing distance from the seaface.

Tidal marshes of the Thakuran Basin are characterised by natural halophytic vegetation and dwarf mangrove bushes in the upper intertidal to supratidal regions, which are important for understanding the hydrodynamics of marsh sedimentation and the role of vegetation as sediment baffles. They specify different environmental zones marginal to the creek. The mangrove swamps marginal to the river basin migrate with tidal rhythms and perform a vital role in binding the intertidal sediments. Both marsh and swamp vegetation play a major role in sediment baffling and accretion.

Depositional features typical of a tidal environment such as mud couplets, tidal bedding, and tidal bundles have been recognised. Both ebb-flood and neap-spring cycles have been established from the preserved physical sedimentary structures. A general increase in the thickness of cross-bedded and laminated sets with an increase of tidal amplitude has been observed. Macrobenthic animals include various species of gastropods, pelecypods, crustaceans, polychaetes, and fishes. They play a vital role in the formation of several surface and internal bioturbation structures which can be further used for identifying specific geomorphic zones. Supratidal and intertidal depositional features of the Thakuran River have been clearly separated. Mineral suits including both light and heavy fractions have been examined to infer the acidic igneous and metamorphic provenance. The assemblage of heavy minerals refers to a major Himalayan source derivation of the sediments together with contributions from the Precambrian terrains and Pleistocene terraces. It is expected that the findings of this study would be a potential contribution in the field of tidal sedimentation of the depositional environment of estuarine and coastal Sundarbans of eastern India.

Kolkata, India Gautam Kumar Das

Acknowledgements

I feel honoured in recording my deep sense of gratitude to Prof. Asokkumar Bhattacharya, former head, Department of Marine Science, Calcutta University, who provided constant guidance and supervision during this study. The author gratefully acknowledges various constructive comments and fruitful suggestions offered by Sri S. Rakshit, ex-director, Geological Survey of India, Dr. Manjari Bhattacharji, Associate Professor of Geography, Viswabharati, Santiniketan, West Bengal, and Prof. Siddhartha Datta, Former Pro-Vice Chancellor of Jadavpur University. The author acknowledges the help rendered by the directors of Indian Meteorological Department (IMD), Geological Survey of India (GSI), River Research Institute (RRI), and Zoological Survey of India (ZSI). Titas helped in drawings and corrected the proofs.

Contents

1 Thakuran Drainage Basin ... 1
1.1 Tectonic Set up ... 2
1.2 Geological Framework ... 2
1.3 The Thakuran Drainage Basin ... 3
1.4 Braiding of the River ... 5
1.5 Formation of Islands ... 7
1.6 Summary ... 7
References ... 7

2 Climate and Hydrography ... 9
2.1 Wind, Temperature and Rainfall ... 10
2.2 Tropical Cyclones ... 11
2.3 Tides ... 11
2.4 Tidal Flow Regime ... 14
2.5 Summary ... 15
References ... 16

3 Geomorphic Environments ... 17
3.1 Flow Patterns ... 18
3.2 Geomorphic Environments ... 18
3.3 Geomorphic Divisions ... 18
3.3.1 Description of the Geomorphic Divisions ... 19
3.4 Sand-Body Geometry of the Geomorphic Zone ... 31
3.5 Summary ... 32
References ... 32

4 Sediment Composition ... 35
4.1 Lithogenic and Biogenic Components ... 36
4.2 Source Rock of River Sediments ... 39
4.3 Summary ... 40
References ... 40

5 Mudflats and Tidal Creek Sedimentation . . . 41
5.1 Physical Parameters of Muddy Sediments . . . 42
5.2 Sedimentation Types on Mudflats. . . . 44
5.3 Sedimentation on a Tidal Creek . . . 44
5.3.1 Mud Ridges . . . 45
5.3.2 Mud Microdelta . . . 45
5.3.3 Mud Pellets and Mud Lumps . . . 46
5.3.4 Wood Clumps on Mudflats . . . 47
5.4 Stratification in Tidal Creek Deposits . . . 48
5.4.1 Epsilon Cross-Stratification (ECS). . . . 48
5.4.2 Alternate Sand-Mud Sequence (Heterolithic Sequence). . . . 48
5.4.3 Convolute Lamination . . . 49
5.5 Summary . . . 50
References. . . . 50

6 Sediment Texture . . . 53
6.1 Shape of the Cumulative Curves . . . 54
6.1.1 *Type I*: Cumulative Curves Having Size Range Within 0.0–4.0 phi, i.e. Within Sand Sizes . . . 54
6.1.2 *Type II*: Cumulative Curves Having Size Range Within 2.5–9.5 phi, i.e. Ranging from Fine-Sand to Clay Sizes. . . . 55
6.1.3 *Type III*: Cumulative Curves Having Size Range −1.0–9.0 phi, i.e. Ranging from Very Coarse-Sand to Clay Sizes. . . . 63
6.2 Inter-relationship Between Grain Size and Morphological Units. . . . 66
6.3 Classification of Sediment Types . . . 66
6.4 Sediment-Trend Matrix of Grain-Size Relations . . . 72
6.5 Sediment—Hydrodynamic Relation . . . 73
6.6 A Bipartite Model of Grain-Size Distribution . . . 74
6.7 Interrelationship of Grain-Size Parameters and River Distance . . . 76
6.8 Environment Sensitiveness of Size Parameters . . . 78
6.9 Summary . . . 79
References. . . . 80

7 Sedimentary Structures. . . . 83
7.1 Bedforms Characters. . . . 84
7.2 Sedimentary Structures on the Mid-Channel Bars. . . . 84
7.2.1 Megaripples. . . . 84
7.2.2 Sandwaves . . . 87
7.2.3 Internal Physical Structures of the Mid-Channel Bars . . . 89
7.2.4 Modification-Features Resultant from Unsteady Flow in Tidal Environment . . . 95

7.2.5 Description of Small-Scale Bedforms 98
7.2.6 Some Characteristics Ripple Types and Other Structures of Tidal Origin 99
7.3 Sedimentary Structures of the Point Bars 107
7.3.1 Description of Bedforms 107
7.3.2 Description of Internal Structures 108
7.4 Sedimentary Structures of the Swash Platform 110
7.4.1 Description of Bedforms 110
7.5 Sedimentary Structures of Wash-Over Flats 113
7.5.1 Surface Features 113
7.5.2 Internal Sedimentary Structures of the Wash-Over Flat. . . . 114
7.6 Remarks on Channel Configuration and Channel-Fill 118
7.7 Summary 118
References. 119

8 Bioturbation Structures 123
8.1 Basis of Classification of Bioturbation Structures 124
8.2 The Classification 128
8.2.1 Surface Bioturbation Structures 128
8.2.2 Internal Bioturbation Structure 134
8.2.3 Dwelling Structures: Burrows and Tubes Made by Organisms in Which They Live or Use for Escape 135
8.3 Summary 139
References. 139

9 Mangroves Swamp and Tidal-Marsh Sedimentation 141
9.1 Geomorphological Features of the Mangroves and Marshes 142
9.2 Mangrove Characteristics 142
9.3 Tidal Sedimentation on Mangrove Swamps 143
9.4 Tidal Marshes. 147
9.5 Intertidal Marsh Sedimentation 149
9.6 Texture of the Marsh Sediments 149
9.7 Physical and Biogenic Structures 150
9.8 Summary 151
References. 151

Abbreviations and Units

DO	Dissolved oxygen
IMD	Indian Meteorological Department
LOI	Loss on ignition
MAB	Man and Biosphere
ppm	Parts per million
ppt	Parts per thousand (‰)
ROOM	Readily oxidisable organic matter
SBR	Sundarbans Biosphere Reserve
STR	Sundarbans Tiger Reserve
TDS	Total dissolved solids

Chapter 1
Thakuran Drainage Basin

Abstract The most conspicuous geomorphic feature of the Sunderbans is a low, alluvial plain intersected by an intricate network of tidal rivers, creeks, estuaries and minor tidal-water courses. The Thakuran River is one of the most important tidal rivers of a meso-macrotidal regime having no perennial freshwater supply, and it forms a drainage basin covering about 900 km^2 in area. The meandering and bifurcating channel systems of the Thakuran Basin have given rise to the formations of various islands mainly by the settling down of particles on the river bed and also by being cutting off from the mainland due to the bifurcation of channels. Construction of embankments to prevent severe bank erosion has resulted in the changing courses of the Thakuran River and the hindrance to growth of mangroves along its river banks.

Keywords Geological framework · Tectonic set up · Braiding of the river · Formation of island · Meso-macrotidal setting · Embankments · Thakuran river · Sunderbans

The deltaic Sunderbans of eastern India are characterised by low, alluvial plains covered by natural mangroves swamps and marshes and are intersected by a large number of tidal rivers, tidal inlets and creeks, estuaries and a network of intricate minor saltwater courses. Many of these serve dual purposes of fresh-water discharge and the to-and-fro movement of the saline water wedge arising from the sea surface. The coastline of West Bengal is comprised of the tidal plains of a large number of estuaries and tidal inlets of which the Ganges-Brahmaputra tidal plain is the most important. Tidal flooding, river fluxes, waves, sea level change, regular episodic and non-episodic events like cyclones and storms are the dominant physical processes that continually modify this coast over the short- and long-term time scales.

The drainage basin of the Thakuran River (latitude 21°35′–22°10′ N and longitude 88°25′–88°35′ E), one of the many tidal water courses, is bounded by the Matla and the Saptamukhi Rivers in the east and the west respectively. In the strict sense, the Thakuran River, like many other so called rivers, viz. the Saptamukhi, the Matla etc., is a large tidal creek in this low-lying coastal plain. It has no perennial fresh-water source. The river is the result of tidal incursion and retreat. During this

G.K. Das, *Tidal Sedimentation of the Sunderban's Thakuran Basin*,
DOI 10.1007/978-3-319-44191-7_1

process of to-and-fro movement of tidal water on a gently sloping plain, it is connected by many other creeks of lesser and lesser magnitude variously called locally as ‘khal’, ‘gang’ etc. The orientation, geomorphic setting and materials of the river deposits are resultant products of the complex interaction of several factors like the tectonic framework of the Bengal Basin, and geological, climatological, physical, chemical and biological processes. The geographical location of the river also acts as an important parameter as all other factors are subjected to change depending on its location. The morphodynamic changes of the coastline are assumed to be the outcome of: (i) long-term events like tectonic and geological processes and (ii) short-term events like tides and waves. Both these events act jointly in this highly-dynamic macrotidal coastline.

The river mouth widens into a funnel as it approaches the Bay of Bengal. This feature of the river, like many others, characterises this coast of West Bengal that causes its designation as a deltaic-estuarine coast. In all, there are twelve major tidal rivers, plus estuaries and many minor inlets and creeks, which have given rise to the much—indented nature of the coastal Sunderbans.

1.1 Tectonic Set up

The coastline of West Bengal belongs to the Amero trailing-edge coast (Inman and Nordstrom 1971; Davies 1972), whereas, the opposite continental coast is a collision coast. This plate-tectonics-related factor predominately control the evolution of the coastline of West Bengal. The present-day drainage pattern of this coastal area results from two distinct factors noticeable to-date throughout the Tertiary history of the Bengal Basin (Biswas 1963; Sengupta 1966). These are: (i) a regional southerly slope of the Bengal Basin due to the movement on the hinge zone at the edge of the shelf and (ii) the increasing rate of southerly tilt of the West Bengal part of the Bengal Basin due to a relatively greater rate of subsidence of the south-western part of the hinge. This tectonic control of the basin is further accentuated due to the presence of numerous faults, which have been intermittently active during the Quaternaries (Morgan 1970). The Thakuran River, along with all other river systems on the plains of the Bengal Basin, maintains a north to south trend and reflects the surficial manifestation of the subsurface southerly tilt of the Bengal Basin as a whole. This southerly tilt, however, has been thought to have been more effective right from the late seventeenth century (Sengupta 1966, 1972).

1.2 Geological Framework

Being situated on the delta of the Ganges and Brahmaputra Rivers, the depositional behaviour of the river is primarily controlled by the regional geology of delta formation involving both fluvial and coastal processes (Fig. 1.1). The deltaic zone

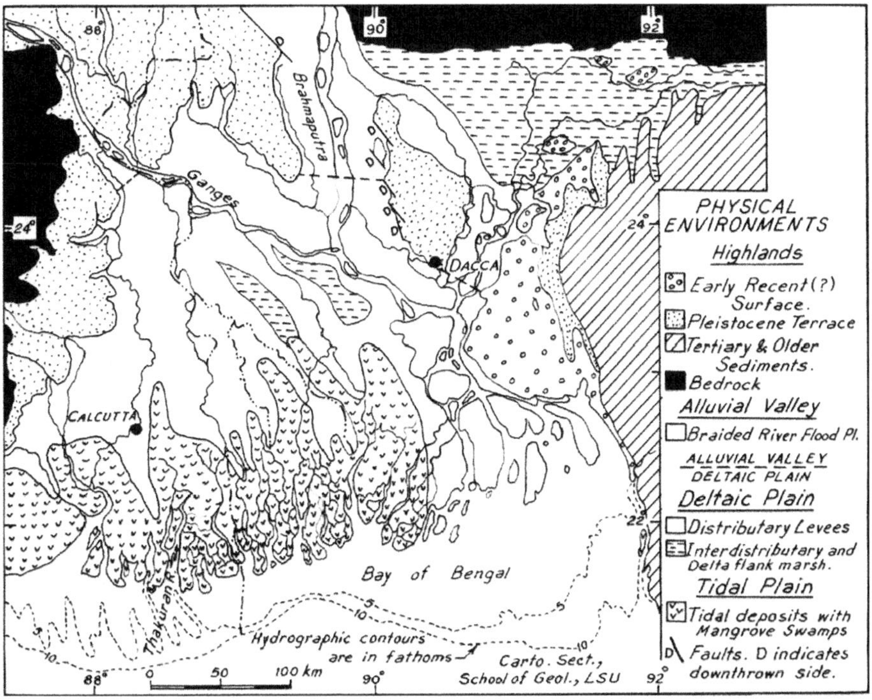

Fig. 1.1 River Thakuran in the depositional environments of the Ganges-Brahmaputradelta covering India and Bangladesh

represents a down-warped basin flanked by successively occurring older rocks from Tertiary and older sediments to younger Pleistocene Terraces into which the major river valleys have been incised (Morgan 1970). High macrotidal amplitudes (4.6–5.5 m spring tide) along the coast results in strong tidal currents leading to a network of deeply scoured (>30 m) tidal rivers and inlets and the formation of an overlapping tidal plain across the intermittently subsiding Bengal delta (Morgan 1970). The Thakuran River is one such tidally-scoured channel having a present-day thalweg level ranging from −10 to −20 m, particularly from its middle to the seaward stretches.

1.3 The Thakuran Drainage Basin

The Thakuran River occurs as an important drainage basin over the basement of alluvial and tidal sediments. It has a length of 80 km from north to south. From a very small width of less than 1 km at its northern uppermost part, the funnelled mouth of the river attains a width of about 8 km before meeting the Bay of Bengal in the south. The river is entirely tidal and belongs to a meso-macrotidal regime from its mouth (5.5–4.6 m spring tide) further upstream (3.2–2.8 m spring tide). The basin occupies approximately 900 km^2 area and is characterised by a number of small islands and sediment bars separated from one another by numerous tidal water-ways. It belongs to southerly sloping lowland and is flanked by fringing mangroves swamps all along the banks covering the inter-channel areas.

The Thakuran River is not directly fed by any freshwater from its landward side like the Ganges. Rainwater and run-off water from the catchment are its only sources of freshwater. Thus, the decrease in salinity of the Thakuran River during

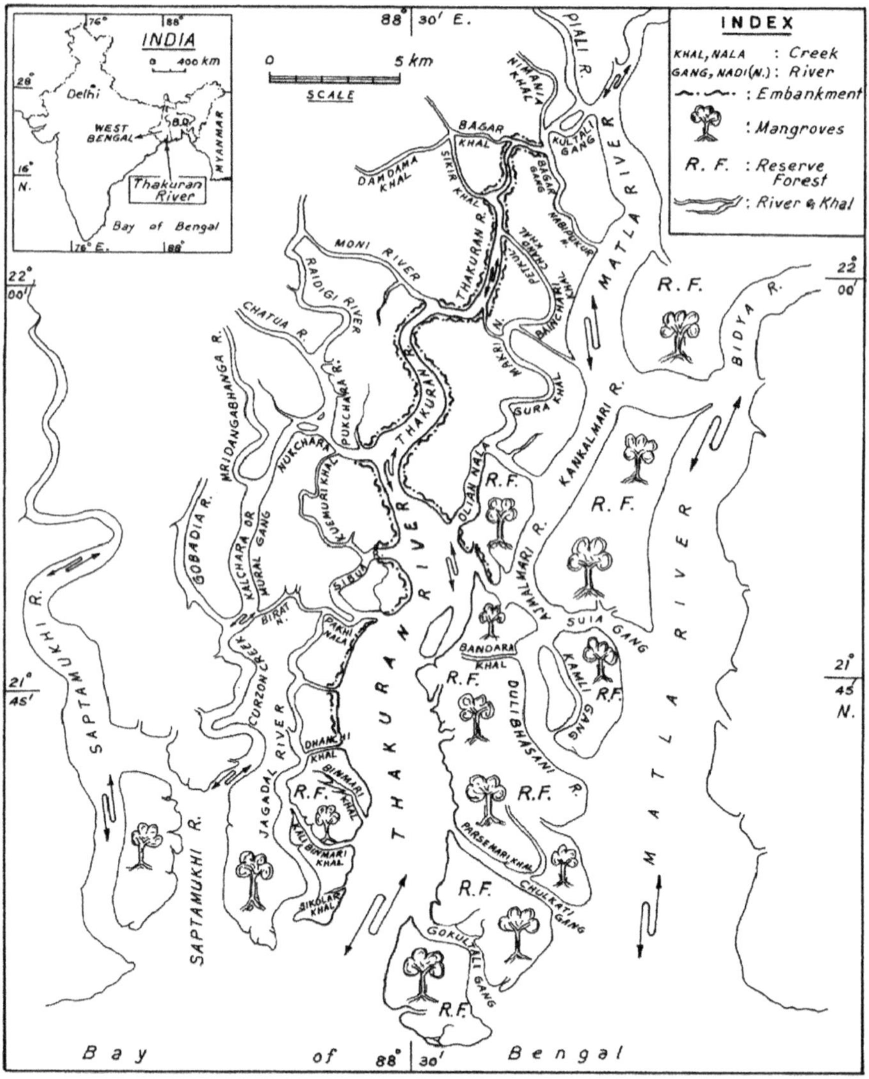

Fig. 1.2 River Thakuran with its connective drainage network and embankments

monsoon reflects the effect of high monsoonal precipitation and subsequent run-off of rainwater into the river basin. The middle stretch of the river is bifurcated into two separate channels, the western smaller channel known as Jagadal gang (river) and the main flow in the east as the Thakuran River (Fig. 1.2). Both these bifurcated channels have a more or less parallel north-south trend. The Jagadal gang at its upper end is connected to the Thakuran River by the Chirapat or Pakhirala creek. Beyond this point of bifurcation further north, the Thakuran gradually becomes thinner and ultimately connects the Matla River through the Nabipukur-Piali, a distributary of the Matla. At this point, there is an outfall of another inland channel called the Nimania Khal (creek) from the northwest.

1.4 Braiding of the River

The Thakuran River shows a braided course particularly in its middle stretch. It has a winding course before its connection with the Matla River to its east. The Thakuran is also well-connected with the Saptamukhi River to its west through the meandering Kalchera-Kurzon creek. Thus the Thakuran is located more or less centrally between Matla and Saptamukhi and forms a complex maze with numerous interconnecting tidal-spill channels. In the process of formation of this intricate maze, the meandering and bifurcating channel systems enclose large and small land blocks in the form of islands. Furthermore, the deposition of sand bars on the river bed also helps in braiding the course of river.

The Paschim Sripatinagar Island, which is 40 km downstream, the Upendranagar island (48 km downstream) and the Dhanchi Island (65 km downstream) are areas of the mainland separated by the encircling courses of the Sibua River, Jgadal River and Kali Binmari Khal respectively. In contrast to the above process of island formation, the mid-channel bars of the Paschim Sripatinagar, Sridharnagar and of many other are products of bedload deposition on the river-bed itself.

The Paschim Sripatinagar and the Sridharnagar mid-channel bars are gradually gaining stability by the establishment of mangrove saplings like *Avicennia marina, A. alba, A. Officinalis*, and *Sonneratia apetala* and sea grass like *Porteresia coarctata*. The erosional margins of these islands exhibit complete sections of the intertidal sedimentary facies. The small insignificant mid-channel bars are generally free from vegetation. They are unstable and migrate seasonally with the tidal currents. Top surfaces of these bars are exposed only during low tides. Trench sections through these bars appear useful to differentiate subtidal and lower intertidal facies. But it was difficult to examine such sections because of high-water saturation leading to quick collapse-of the trench walls.

Following a different mechanisms, the mid-channel islands of the Thakuran River have evolved from the point-bar deposits. The bank-ward edge of the point bars may be cut off by some spill channels emerging from the main river. This disconnects the point bars from the adjacent banks. The isolated portion of the point bars then rest on the main river bed in the form of mid-channel bars (Das 2015). The formation of Bhubaneswari Island and Damkal dwip in the upper stretch of the Thakuran can be explained in this way. Natural cliffs marginal to these islands exhibit deposition from intertidal to supratidal facies (Klein 1971; Terwindt 1988). The Bulchery Island at the mouth of Thakuran represents an ebb-tidal delta (bar) and is located between the flood channel of the Thakuran River and the ebb channel of Thakuran known as the Gokultali Gang (Fig. 1.3).

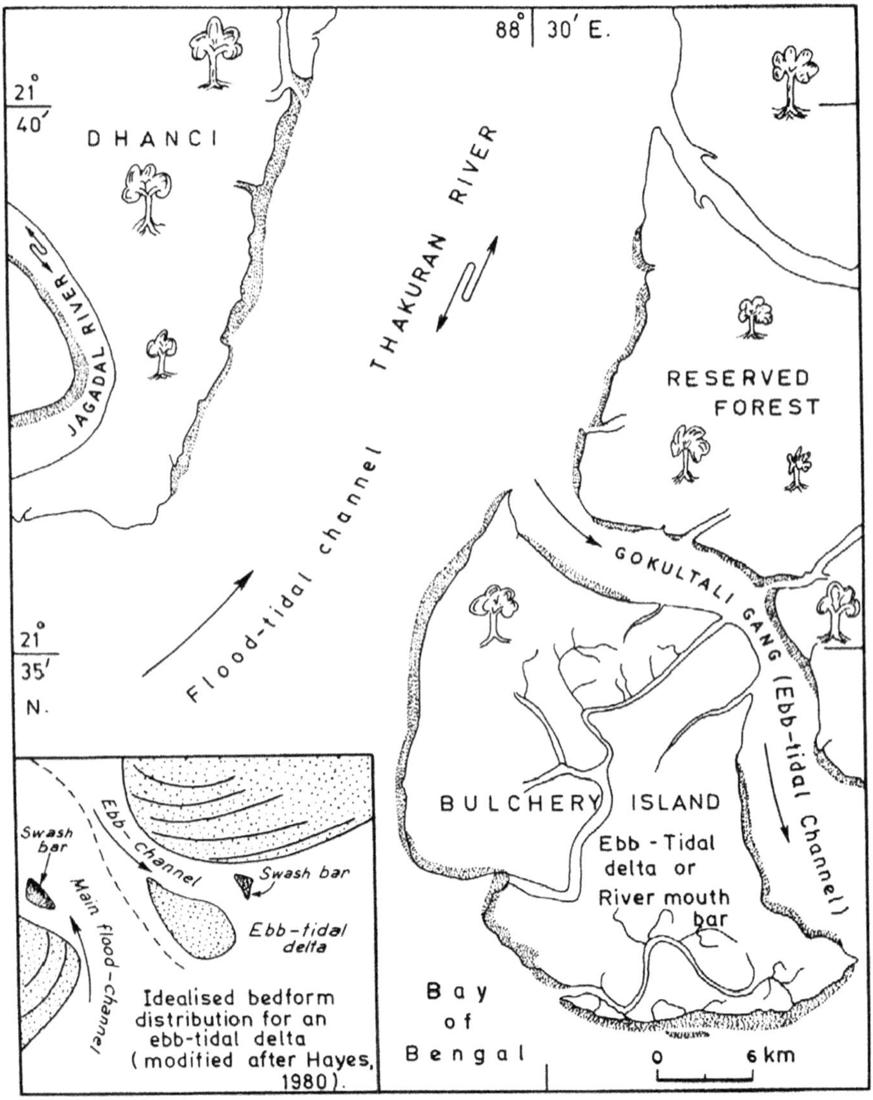

Fig. 1.3 Map of Thakuran river mouth showing ebb-tidal delta or river mouth bar formation. An idealized bedform distribution for an ebb-tidal delta is shown in *inset*

1.5 Formation of Islands

There are three chief mechanisms of island formations noticeable in the Thakuran River basin:

(i) depositional islands originated from the deposition particles on the riverbed to form the mid-channel bars and the channel islands. Having been formed primarily under the influence of flood currents, these flood-tidal bars have the tendency to move upstream. The river-mouth island (bar) formed by the influence of ebb currents, however, has the tendency to move seaward;
(ii) mid-channel bars evolved by disconnection of the bank-ward edge of the point bar. The bank-ward edge of the bar is occupied by active or abandoned spill channels; and
(iii) islands formed by mainland cut off during bifurcation of channels resulting from shallowing of river beds. This is associated with intermittent subsidence and sedimentation of the Bengal delta (Morgan 1970).

1.6 Summary

Numerous low, swampy areas in the Thakuran River occur in the topographic depression between old and new channels. The Sunderbans with its numerous island systems, may thus be described as a group of islands separated by saline watercourses. The intricate tidal pathways, islands and swamps in the physiographic complexion of the Thakuran River are continuously changing over space and time. The present network of land and water undergoes evermore modification and has undergone similar changes in the past. With this perspective of the past and present configuration of the land-water network, the future trend of change in the configuration of the rivers of the Sunderbans including Thakuran may be understood.

References

Biswas B (1963) Results of exploration for petroleum in the western part of Bengal basin, India. Proc 2nd Symp Dev Ptr Res ECAFE Min Res Div Ser 18(1):241–250

Das GK (2015) Estuarine morphodynamics of the Sunderbans. Coastal research library, vol.11. Springer, Switzerland, 211p

Davies JL (1972) Geographical variation in coastal development. Oliver and Boyd, Edinburgh, 204p

Inman DL, Nordstrom CE (1971) On the tectonic and morphologic classification of coasts. J Geol 79:1–21

Klein GDV (1971) A sedimentary model for determining palaeotidal range. Geol Soc Am Bull 82:2585–2592

Morgan JP (ed) (1970) Deltaic sedimentation: modern and ancient. Spl Publ Soc Econ Paleon Miner 15:312p, Tulsa

Sengupta S (1966) Geological and geophysical studies in Western part of Bengal basin, India. Bull Amer Asso Petrol Geol 50:1001–1017

Sengupta S (1972) Geological framework of the Bhagirathi-Hugli basin. Proceedings of the interdisciplinary symposium. Calcutta University Publ, pp 3–8

Terwindt JHJ (1988) Paleo-tidal reconstruction of inshore tidal depositional environments (pp 233–264). In: deBoer PL, VanGelder A, Nio, SD (eds) Tide-influenced sedimentary environments and facies. D Reidel Publishing Company, 530p

Chapter 2
Climate and Hydrography

Abstract The Thakuran Basin is an ideal example of a region with a tropical climate with hot, rainy and humid summers and dry winters. The monsoon is normally characterised by the presence of cyclonic storms which bring large-scale littoral drift and coastal modifications. The Thakuran River belongs to a meso-macrotidal regime with semidiurnal tides for its entire 80 km length. Separation of flood and ebb channels in the river basin is controlled by the principle of Coriolis force. Time-velocity asymmetry of the tidal current is appreciable and marked by bedform configurations related to flood and ebb channels.

Keywords Tide · Rainfall · Temperature · Salinity · pH · Wind velocity · Semidiurnal tide · Meso-macrotidal regime · Froude number · Reynolds number · Thakuran river · Sunderbans

The climate of the Sunderbans including the Thakuran basin is tropical oceanic. Three seasons viz., winter (November to February), summer (March to June) and monsoon (July to October) are easily recognizable. Winter temperature ranges from 10 to 25 °C and summer temperature from 28 to 36 °C. The annual rainfall ranges between 1470 and 2210 mm. Salinity of coastal water ranges between 22 and 31 ‰, while becoming less in the estuaries. The pH of coastal water ranges from 7.5 to 8.5. The generally mesotidal coast (tidal amplitude 2–4 m) is macrotidal (tidal amplitude >4 m) at the funnel mouths of estuaries and large rivers. The tides are semi-diurnal with slight diurnal inequality (Das 2015). The heavy rainfall during the monsoon influences the tidal interactions in almost all the rivers of the Sunderbans, and flood- and ebb-tidal currents fluctuate with the seasons. The maximum wind velocity is 16.7–50 km/h (April to June) and the minimum wind velocity is 10.7–11.8 km/h (December to February). West Bengal is cyclone prone, with three to four severe cyclones per year. The wind velocity during cyclones often ranges from 80 to 140 km/h. Cyclones also initiate large-scale littoral drift and lead to devastating coastal modifications. Wave heights range from 0 to 0.6 m with a

G.K. Das, *Tidal Sedimentation of the Sunderban's Thakuran Basin*,
DOI 10.1007/978-3-319-44191-7_2

wave period of 5–7 s during the calm winter season, whereas, these become 1.8–2.4 m and 12–14 s respectively during the rough summer seasons. Wave height can exceed far above 2.5 m in the event of cyclonic storms.

2.1 Wind, Temperature and Rainfall

Three climatic seasons of a year are recognisable in the Hugli-Matla estuaries of the Sunderbans. These are: (i) dry season (pre-monsoon) Feb to May, (ii) rainy season (monsoon) June to September; and (iii) winter season (post monsoon)—October to January. Seasonal variations of wind velocity, wind direction and rainfall during pre-monsoon, monsoon and post-monsoon are quite significant in this tropical area.

The mean wind velocities during the three principal seasons, i.e. pre-monsoon, monsoon and post-monsoon, are 11.5, 11.1 and 6.65 km h^{-1} respectively. The south-southwest to southwest wind direction of pre-monsoon and monsoon changes to north-north east to northeast during the post–monsoon times. The percentage of rainy days sharply rises to 65.79 % during monsoon, from the pre-monsoonal 20.35 %, and again declines to 14.81 % during the post–monsoon times.

Rainfall data measures a mean annual value of 1908.4 mm. This became slightly higher to the tune of 1949.4 mm during the years 1991–1993. The three-year

Table 2.1 Seasonal variations of water salinity and water temperature in the Thakuran River

Parameters	Pre-monsoon	Monsoon	Post-monsoon
Water Salinity (‰)	16.8–28.4	8.6–13.6	15.1–23.6
Water temperature (°C)	28.4–31.7	28.3–29.9	25.3–26.9

Table 2.2 Variations of salinity and pH along the stretch of the Thakuran River

Location	Salinity (ppt)	pH
Jata	15	7.81
Bhubaneswari	16	7.77
Saheber Ghat	17	7.73
Nandir Ghat (Maipith)	21	7.81
Harinala khal	20	7.53
P. Sripatinagar	20	7.74
Chilkamari	21	7.63
Hazarbigha	21	7.65
Lakshmi Janardanpur	21	7.15
Upendranagar	21	7.91
Rakhalpur	22	7.85
Sridharnagar	23	7.88
Dhanchi II	21	7.69
Dhanchi III	20	7.66
Dhanchi IV	23	7.23
Dhanchi South	23	7.61

average (1991–93) of evaporation data was recorded as 977 mm, and this implies an excess of annual precipitation in the area over annual evaporation.

Table 2.1 presents salinity and temperature variations in the waters of the Thakuran River. Salinity varies from 8.6 to 28.4 ppt from monsoon to pre- and post-monsoon times (Table 2.1). On the contrary, the salinity of open seawater off the mouth of the rivers varies between 23 and 31 ppt. The pH of the river water varies between 7.15 and 7.91 (Table 2.2).

2.2 Tropical Cyclones

Tropical cyclones with variable wind speeds (63–87 km h^{-1}) are regular phenomena in the Sunderbans area. "Tropical cyclonic depressions" with wind speed <63 km h^{-1} pass over the area during July to September. "Cyclonic storms" with wind speed ranging from 63 to 87 km h^{-1} strike the area during June to September. Most of destructive "severe cyclones" with wind speed >87 km h^{-1} affect the area particularly during May and September. The average occurrences of cyclonic depressions, cyclonic storms and severe cyclonic storms over the study area are twice a year, once every three years and again once every three years, respectively. The 2009 cyclone that crossed Sunderbans on 25th May was the most recent major event until this writing. Its maximum surface wind speed reached 130–148 km h^{-1}. This storm caused wide-spread damage in the coastal area of West Bengal and the adjoining country Bangladesh.

2.3 Tides

The entire coastline of West Bengal with its highly indented nature is the outcome of differential erosion and accretion caused by fluvio-tidal and tidal processes in an estuarine delta framework. The perpendicular-to-shore orientation and elongated shape of most of coastal islands reflects the impacts of flood- and ebb-tidal currents in a macrotidal coastal environment (Davies 1972). The intertidal landforms also result from high tidal fluctuations with unequal inundations and exposures.

The Thakuran River is tidal for its entire 80 km length and belongs to the meso-macrotidal setting. At the sea face, the mean tide is 5.5 m whereas, in the middle stretch, it declines to 3.5 m. At Sagar Island, the mean maximum spring tidal range varies between 6.5 and 7.5 m and the mean minimum neap between 2.0 and 2.5 m. The highest yearly tides are experienced from August to September, while the lowest occur from February to March.

The tide in the Thakuran River is semi-diurnal with little diurnal inequality (Fig. 2.1). The mean ebb velocity is 58 cm/s, whereas, the mean flood velocity is 48 cm/s. Time velocity asymmetry (Fig. 2.1) controls both bedform configurations and their orientations on the flood and ebb channels. Flood and ebb discharge data,

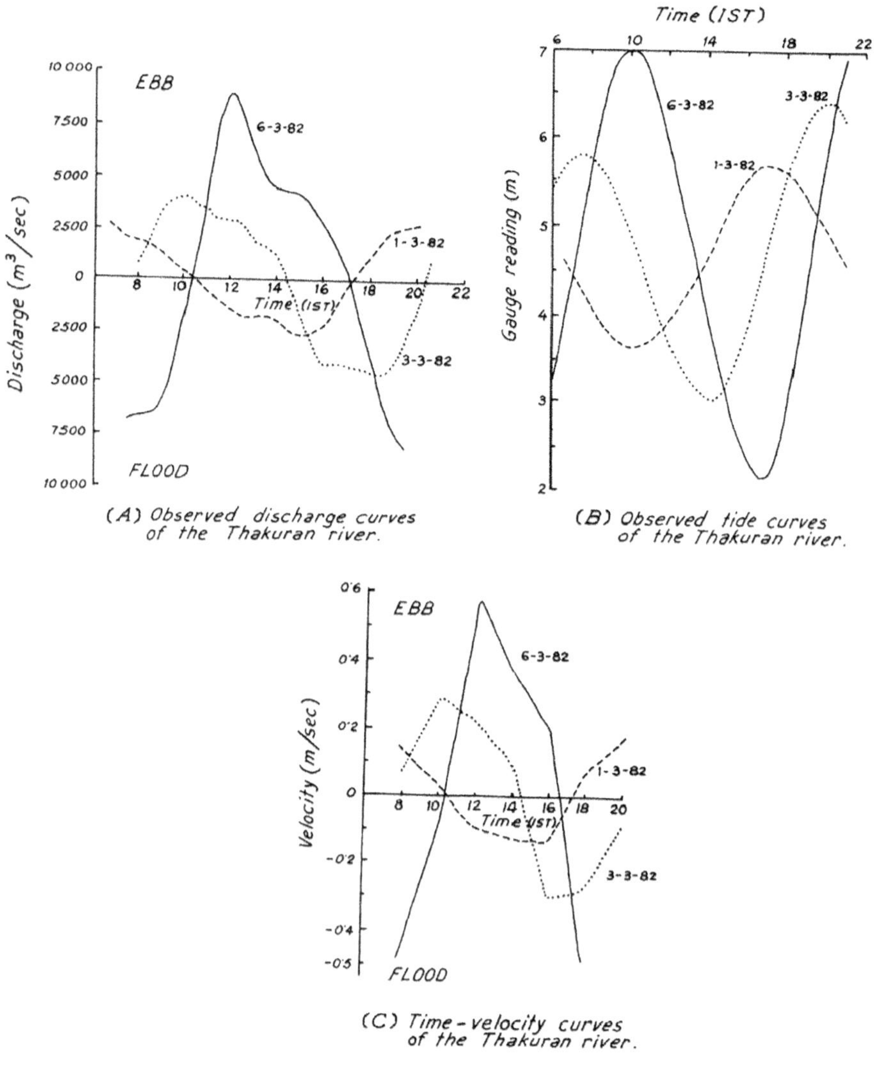

Fig. 2.1 Tidal charts of the Thakuran River

Table 2.3 Variations in ebb-flood discharge, flow velocity and water volume in the Thakuran (RRI, W.B.)

Tide	Discharge (cubic m/s)		Flow velocity (cm/s)		Water volume (millions of cubic metres)	
	Ebb Max	Flood Max	Ebb	Flood	Ebb	Flood
Neap	2400	2600	26	26	36.2	35
Mean	5150	5150	58	48	78.1	71
Spring	8000	7700	91	69	120	107

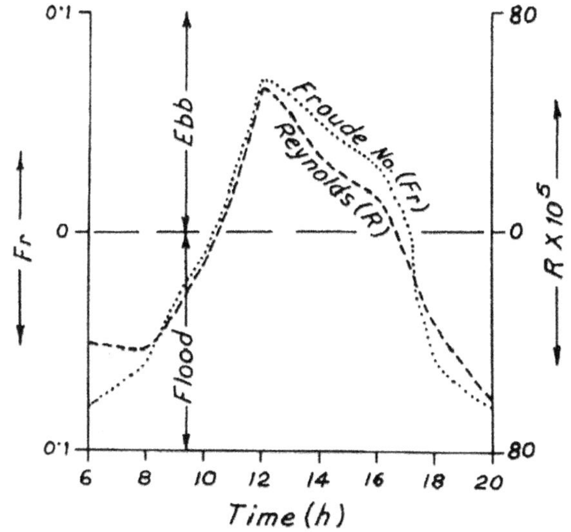

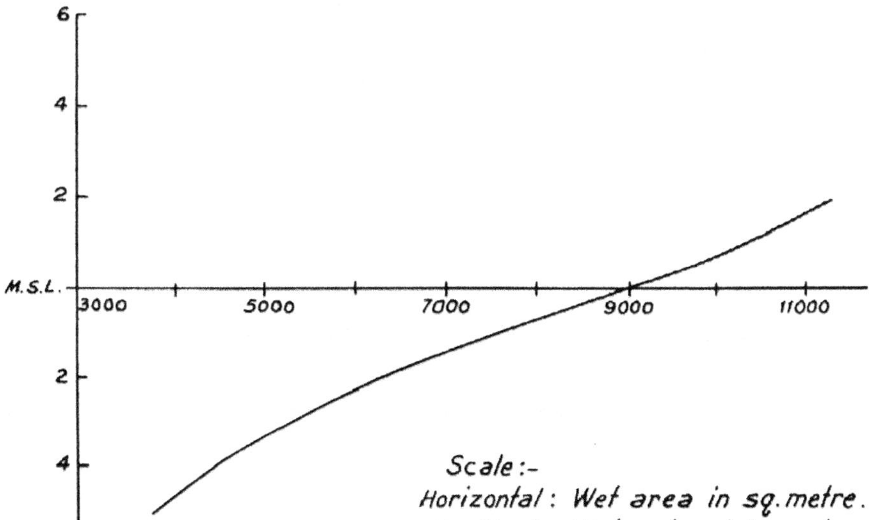

Fig. 2.2 Relation between water level and wet cross–sectional area of the Thakuran River

variations in ebb and flood velocities, and variations in ebb and flood volumes of the Thakuran River in Sunderbans, as obtained from the River Research Institute, Govt. of West Bengal, are summarized in Table 2.3.

The eastern margin of the riverbed is generally dominated by ebb flows, whereas, flood flows dominate along its right-margin thalweg. The ebb and flood channels of the Hugli River below Diamond Harbour also follow the same pattern

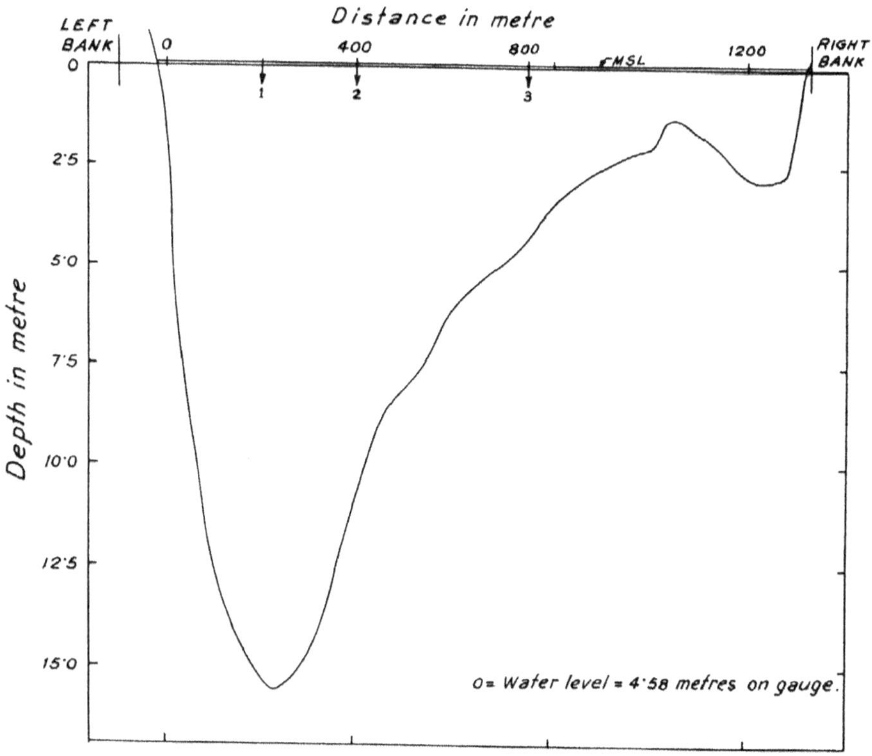

Fig. 2.3 Cross section of the river 40 km upstream from mouth showing asymmetric nature of bottom topography

(Chakraborty and Sen 1972). This separation of flood and ebb channels in the majority of the rivers of coastal West Bengal is hypothesised to take place according to the principle of Coriolis force (Dyer 1986).

Spring tidal data and the corresponding tidal curves of the Thakuran River near Damkal Dwip (40 km upstream) are available from River Research Institute (RRI), Govt. of West Bengal (Fig. 2.1). The relationship between the water level and wet cross-sectional area at this place of the river is shown in Fig. 2.2. The cross section of the river clearly reveals an asymmetric bottom configuration (Fig. 2.3).

2.4 Tidal Flow Regime

The changes in the Froude number (Fr) and Reynolds number (R) corresponding to flood and ebb flows of the Thakuran measured near Damkal Dwip, as a case study, are shown in Fig. 2.2 and Table 2.4. The curves for the Froude number (Fr) and

Table 2.4 Variations of Reynolds number (R) and Froude number (Fr) at a cross section near Damkal Dwip over time in a tidal cycle

Time in hours	Reynolds number (R) (X 10^5)	Froude number (Fr)
6	43.8	0.08
8	45.8	0.06
10	11.01	0.01
12	56.73	0.07
14	27.0	0.05
16	13.84	0.03
18	33.72	0.06
20	67.14	0.08

Reynolds number (R) exhibit a close similarity in pattern and designate a lower flow regime in which only ripples of different scales are generated. The calculated value of the Reynolds number (R) from the same area (Table 2.4) reveals that the flow, except for a short period, is turbulent almost throughout the tidal cycle. The term unsteady flow implies that the velocity at a point varies with time. Non-uniform current signifies variations of velocity in space, and turbulence means random variation of instantaneous velocity with respect to both time and space about some mean value.

2.5 Summary

The Thakuran River bed of Sunderbans offers a unique opportunity for the study of unsteady flow in natural settings. The unsteady flow conditions of the river generally arise from: (i) the fluctuations of depth and velocity within ebb and flood periods; (ii) flow reversals associated with ebb and flood periods; and (iii) variations between neap and spring tidal flow. There may be even longer cycles to control such conditions. Thus, many of the bedforms cannot be explained with the classical flow-regime concept valid for the steady-flow conditions (Simons and Richardson 1962; Harms and Fahenstock 1964).

Bedforms like ‘skewed spurs’, ripple fan, planed-off crests of megaripples and micro-deltas are some of modification features arising from fluctuating velocities and variable depths within ebb and flood periods. Reversals of flow lead to modifications of bedforms where flood–oriented megaripples exhibit slight rounding of crests accompanied by small asymmetric ebb-oriented aprons (Elliot and Gardiner 1981; Bhattacharya 1993). Decaying megaripples and partly reversed and asymmetrical megaripples reflect modifications due to the contrast between neap and spring levels.

References

Bhattacharya A (1993) On limitations of flume experiments, concept of flow regime and related bedforms in sedimentology, 50 years commen vol River Research Institute, West Bengal, pp 68–69

Chakraborty AK, Sen A (1972) Sediment transport characteristics in the Hugli and the effects of upland discharge. In: Bagchi KG (ed) The Bhagirathi-Hugli Basin, Calcutta University Publications, pp 89–90 (373p)

Das GK (2015) Estuarine morphodynamics of the Sunderbans. Coastal Research Library, vol 11. Springer, Switzerland, 211p

Davies JL (1972) Geographical variation in coastal development. Oliver and Boyd, Edinburgh 204p

Dyer KR (1986) Coastal estuarine sediment dynamics. Wiley, New York, 342p

Elliot T, Gardiner AR (1981) Ripple, megaripple and sandwave bedforms in the macrotidal Loughor Estuary, South Wales, UK. Spl Publ Int Asso Sed 5:51–64

Harms JC, Fahenstock RK (1964) Stratifications, bedforms and flow phenomena. Bull Am Asso Petrol Geol 48:530–544

Simons DB, Richardson EV (1962) Resistance to flow in alluvial channels. Am Soc Civil Engrs Trans 127:927–953

Chapter 3
Geomorphic Environments

Abstract The most recognisable geomorphic environments of the Thakuran Basin are river systems of fluvial, estuarine and tidal nature, tidal creek systems and sandy delta beaches. The roles played by these dynamic agencies have given rise to various erosional and accretional features throughout the drainage basin. Natural levees, point bars, mid-channel bars, swash platforms, wash-over flats and an ebb-tidal delta are some of the noteworthy geomorphic features present in the area. Severe bank erosion is observed in areas of high wave action near the mouth of the Thakuran River. Point bars are generally fulcate to crescentic in shape with Epsilon Cross Stratification. Depending upon the distance from the sea, the mid-channel bars show varying composition from mud to sand. Both banks of the Thakuran River exhibit fringing mangrove vegetation. The distal end of the muddy swash platform supports luxuriant mangroves like *Phoenix paludosa* forests, which is the natural abode for the famous royal Bengal tigers.

Keywords River bank · Natural levee · Mid-channel bar · Point bar · Marginal bar · Swash platform · Ebb-tidal delta · Thakuran river · Sunderbans

The Thakuran River channel is characterised by the point bars, mid-channel bars, swash platforms, wash-over flats and a river mouth bar. Physical sedimentary structures are scattered on the point bar along convex and concave banks and upon the riverbed of the meandering estuarine stretch of the Thakuran River. The upper surface of the tidal shoals at the mid channel exhibits a gradual slope to the order of 12°–4° downstream and is traversed by a complex network of tidal creeks. Trench sections in the sand flats exhibit dominance by flaser and wavy bedding, whereas, lenticular bedding and starved-ripple bedding dominate in the mudflats. Alternations of sand and mud deposits in variable proportions reflect deposition from the bed- and suspended-loads characteristic of tide-dominated, high-velocity ebb or flood currents and the corresponding slack-water condition.

G.K. Das, *Tidal Sedimentation of the Sunderban's Thakuran Basin*,
DOI 10.1007/978-3-319-44191-7_3

3.1 Flow Patterns

The geomorphic environments of Thakuran's meandering drainage system, with its erosional and depositional sites, are the resultant products of two distinct flow patterns: (i) the downstream sinuous flow pattern and (ii) the helicoidal flow pattern. The mid-channel bar in a meandering system is either the result of slackening flow velocity due to channel topography or is a result of isolation of the point bars due to subsequent cut-off by the subsequent deflection of the water flow (Sarkar and Basumallick 1968). In a tidal river, the depositional behaviour of the channels are also very sensitive to the seasonal variation of the tidal behaviour. The mid-channel bars in Thakuran tidal river are mostly resultant from the convergence of flood and ebb flows.

3.2 Geomorphic Environments

A complex network of geomorphic environments characterises the Thakuran tidal basin of the Sunderbans. Both sand flats and mudflats occur at the sea face and depend on their locations on high- and low-energy zones respectively. The tidal Thakuran River is a muddy system except for some sandy flats that occur in the mid-channel bars and sandy swash bars at the mouth of the river. The creeks are absolutely muddy systems. Mangrove swamps occur on the intertidal mudflats of estuaries, creeks and inlets. Narrow marshes occur on the upper-intertidal to the supratidal zones in a sporadic manner.

Three broad geomorphic environments can be recognised in the Thakuran River of the Sunderbans. These are: (i) Fluvial, estuarine and tidal river systems with sand flats and mudflats, (ii) Tidal creek systems with mudflats and (iii) Sandy delta beaches.

3.3 Geomorphic Divisions

Many geomorphic environments of deposition and erosion along and across the Thakuran River basin are noticeable. These are as follows: (i) riverbanks/natural levees, (ii) point bars, (iii) mid-channel bars, (iv) swash platforms, (v) wash-over flat and (vi) ebb-tidal delta.

3.3.1 Description of the Geomorphic Divisions

3.3.1.1 River Banks or Natural Levees

Both banks of the river Thakuran are generally characterised by fringing mangroves. As a result, the nature of stratifications of the banks or that of the natural levee deposits is often heavily disturbed by roots and pneumatophores of mangroves and by the churning effect of burrowing organisms. In places of massive erosion, the banks have been fortified by the construction of embankments. As a result, mangroves are truncated in the intertidal zone. Landward, the intertidal area behind the embankments is occupied by marsh vegetation (Bhattacharya 1999). The top-most portions of the bank deposits register evidence of long exposures and contain desiccation cracks and salt-encrusted surfaces in patches. The following features characterise the supratidal bank deposits:

i. Marshy areas are characterised by marsh lamination with fine obscured laminae of silt and clay profusely disturbed by roots of marsh vegetation.
ii. Rapid thinning of sand laminae and the dominance of mud in the upward direction bank deposits are evident in many bank sections (Fig. 3.1).
iii. Plant remains and plant roots constitute a substantial component of bank deposits, particularly towards their upper portions.
iv. The levee surface sometimes shows salt-encrusted patches due to evaporation of saline water pools formed during over-bank flooding.
v. Suspensional mud deposits on the top-most part of the riverbanks exhibit mud cracks or curly mud cracks that indicate subaerial exposure. Curly mud cracks are often associated with areas having algal matting.
vi. Rain prints are occasionally present on the topmost muddy surface.

Fig. 3.1 Thinning of sand laminae and dominance of mud in the higher level. Knife is 26 cm

3.3.1.2 Natural Levee/River Bank Morphology

Both banks of Thakuran register evidence of local sagging (subsidence) in many instances. These areas of subsidence appear as troughs along the banks to be filled up partially or completely with newly accreting sediments. On the initial stage of compaction, these sediments support mangrove saplings. In addition to subsidence, collapsing of banks is locally common due to over-steepening. Inter-bedding of sand with mud-clast-enriched layers is a common feature. Eventually, some portions of the upper levels of the banks slump down to the lower levels. This leads to a mixing of different stratigraphic levels. The over-steepening of the banks is promoted by loosening of bank materials which, inter alia, is caused by the following important factors: (i) semi-diurnal tidal fluctuations and (ii) hydrodynamic conditions that operate in meandering rivers in the form of helical flows. As a result, the over-steepening of the banks slump down to the lower levels.

In vertical sections the bank, deposits are separable into three units based on colour from the top to the mean low-water line. The bank top is dominated by mangrove species *Phoenix paludosa, Ceriops decandra and Excoecaria agallocha* (Fig. 3.2). The top stratum mud deposits (up to 2 m thickness) are vigorously disturbed by the penetration of mangrove roots up to a depth of 1 m or so. The immediately underlying unit is 2 m thick, yellow mud. The unit (generally of 1.5–2 m thickness) is profusely bioturbated by crab burrows and penetrated by mangrove rootlets. The muddy burrow walls and root-affected mud cliffs display

Fig. 3.2 Thakuran river bank at Dhanchi showing occurrence of various mangrove species. 'Ph' stands for *Phoenix* sp, 'Cr' stands for *Ceriops* sp and 'Ex' for *Excoecaria* sp

clear evidence of oxidation marked by the brown sedimentary colour. The lowermost unit of the banks (up to 2 m thickness) is less affected by bioturbational disturbance. In majority of the cases, the topmost unit has an erosional contact with the underlying intermediate unit, whereas no such erosional contact is recognizable between the intermediate and the lowermost units excepting a distinct colour variation.

In areas of old natural levees with more stable and compact substratum, the mangrove plants attain heights above 5 m. Occasionally, these stable banks with an area of some tens of square metres collapse *en bloc* together with the sturdy mangrove trees. The tree trunks often remain in an upright to slightly tilted position even after a vertical fall of 3 m or more (Fig. 3.3). Bank erosion of significant magnitude is noticed in areas of severe wave actions near the mouth of the Thakuran. The muddy riverbank perches over the sandy platforms gently sloping riverward (Figs. 3.4 and 3.5). The top stratum unit is 0.65 m thick and supports dense mangrove forests on its surface. This greyish yellow unit is profusely traversed by roots of mangroves. The underlying unit is 0.9 m thick, yellow mud with intense bioturbation. Below this, a 0.35 m thick dark brown muddy unit is followed by the bottommost 1.1 m thick greyish mud with more intense bioturbation.

Fig. 3.3 Collapsing of the upper stratigraphic level of the bank containing *Ceriops* sp to the lower level leads to amalgamation of material and colour. Note the upright nature of the vegetation hummock even after a vertical fall of 3–4 m

Fig. 3.4 Severe erosion near mouth of Thakuran at Dhanchi. The bank slopes down into gentle platforms. Note the conspicuous nature of erosional mud ridges

Fig. 3.5 Trains of backwash ripples on sandy swash platform. Supratidal mangrove mud slumps over sandy platform

3.3.1.3 Colour Significance of Stratigraphic Units

Colour of the sedimentary units can be divided into four kinds, viz., (i) greenish yellow, (ii) yellow, (iii) yellowish brown to dark brown and (iv) grey to dark grey. The yellow colour is because of staining by ferric compounds and indicates an oxidizing environment. The different shades of yellow to brown are caused by the variation in the state of oxidation. The grey to dark grey colour implies a reducing environment and reflects a normal colour state of yellow and green. The green colour is the outcome of adsorption of ferrous and chlorophyll-related compounds mostly found in mangrove forests (Pantin 1969; Swift and Boehmer 1972; Owens 1981; Stanley and Hait 2000).

The colour variation has a definite stratigraphic relevance. The greenish-yellow laminated units towards the upper part of the sequence indicate relatively greater oxidation caused by exposure. The grey to dark grey muddy units at depth of 3–4 m, on the contrary, reflect a more reducing environment. Local occurrences of isolated small patches of mixed yellow- and greenish-yellow colour within the grey to dark grey muddy units suggest the foundering of upper units.

3.3.1.4 Point Bar

The river Thakuran is macrotidal from its mouth to 40 km upstream. Thereafter, it is mesotidal (2–4 m tidal range) up to its extreme upstream point where it meets the Matla River. The river is generally sinuous to meandering and is characterised by a limited number of point-bar deposits (Table 3.1) where deposition takes place laterally on the convex sides of the meander bends. The sandy point-bar deposits in the macrotidal stretch closer to the sea are gradually replaced landward by mesotidal, gently–inclined, laterally-accreting deposits identical to large-scale

Table 3.1 Length, breadth and length-breadth ratio of the point bars of Thakuran river

Locality	Length (L) (in km)	Breadth (B) (in km)	L/B
Kishoripur	1.25	0.25	5.0
Madhusudanpur	0.1	0.20	0.50
Madhabpur	0.5	0.63	0.79
Sikirhat	2	0.60	3.30
Bhubankhali	4.5	0.4	11.25
Damkal	3	0.75	4.0
Maipit	4	0.63	6.35
Dakshin Kashinagar	5	0.50	10.0
Upendranagar	0.6	0.85	0.70
Sridharnagar	6	0.63	9.52
Dhanchi	5.5	1.10	5.0
Bulchery	4	0.50	8.0

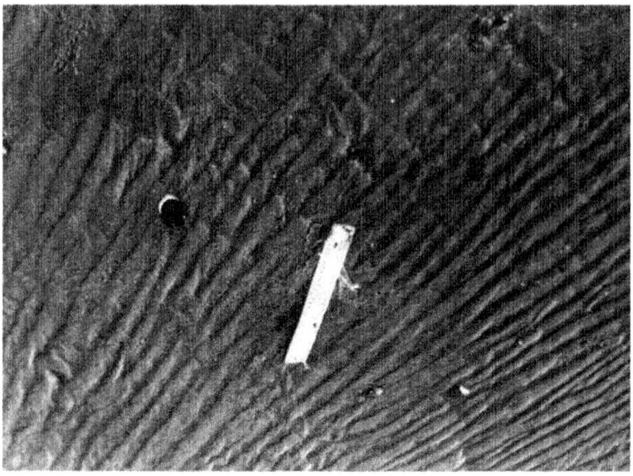

Fig. 3.6 Longitudinal ripples formed by waves and tidal currents. Ripple train is along the direction of current

"epsilon cross stratification" or ECS of Allen (1963) or "longitudinal cross-bedding" of Reineck and Singh (1980).

The point bars of the Thakuran River are generally fulcate to crescentic in shape and extend for 0.1–6 km in length and 0.2–1.1 km in breadth (Table 3.1). Maximum height of point-bars from mean low-water level is variable, 2–4 m. The length-breadth ratio (L/B) of the point bars nicely conforms to that of the meander bends. Sometimes, two to three point bars coalesce along their lengths to give rise to their extremely elongated shapes.

Alternate sand-mud stratifications up to 30 cm thickness are observed on the riverward margins of the point bars at low tide. Trench sections in the point bars also reveal the same pattern of alternation of sand-mud laminae with a gentle riverward dip. Barring localized occurrences of reversing ripples, longitudinal ripples (Fig. 3.6) and wave ripples, the point-bar surface is mostly devoid of any other large-scale bedforms. Human interference very often badly disturbs and obliterates the small-scale bedforms on the point-bar surface.

3.3.1.5 ECS Within Point-Bar Deposits

Large-scale epsilon cross-stratifications (ECS) are gently inclined in the river direction. The stratifications run parallel to the upper surface of the point bars. Thickness of cross-stratified units ranges from 2 to 4 cm. The inclination of the cross-beds averages 15° and varies between 10° and 20°. The ECS of Thakuran River are analogous in dimension to that of the Cretaceous McMurry Formation of the Athabasca oil sands (Rahmani 1988) and the Recent Willapa River point-bar deposits in the USA (Smith 1988).

ECS differs in origin from the cross-stratifications produced by migration of ripples and megaripples (Reineck and Singh 1980). Lateral shifting of point bars facilitates epsilon cross-stratifications to run parallel to the current direction

(Smith 1988). Large-scale ECS are useful for interpreting depositional environments (Oomkens and Terwindt 1960; Van Straaten 1961; Bridges and Leeder 1976; Mossop and Flach 1983).

A muddy substratum supporting mangrove plants overlies the ECS units in many places and this yields a fining upward sequence for the point-bar deposits. Laminations in the muddy units are obscured because of profuse bioturbation and penetration of mangrove roots. Mangroves contribute a substantial quality of organic matter in the uppermost unit. Occasionally mud clasts are trapped within the lower stratigraphic units that are ripped out from the upper stratigraphic horizons. A clear unidirectional current can be inferred from the ECS. Their strike direction is essentially parallel to the riverbanks.

3.3.1.6 Mid-channel Bar

Mid-channel bars are riverbed geomorphic bodies that are quite conspicuous in the central part of the Thakuran River of Sunderbans. These constitute a series of depositional sites on the riverbed itself and are relatively more common from 10 km upstream of the mouth of the river. They occur as isolated, elongated, elliptical, depositional shoals extended along the river-channel axis. The actual extension of the bars and their limits were studied during ebb tides only. The mid-channel bars generally maintain a gradual upward convexity having lengths ranging from 2.1 to 4.75 km and breadth from 1.0 to 3.7 km (Table 3.2). The height ranges from 2 to 4 m above mean low-water level. They migrate upstream with the movement of flood-tidal currents along the main flood channel. At Paschim Sripatinagar, 39 km upstream from the confluence of the river and Bay of Bengal, the rate of upstream migration of a bar has been calculated to be 1.5 km in 90 years, with a rate of 16 m per year (Das 2015). Bedforms and grain-size characteristics of this mid-channel bar have been studied during ebb, when the bar surface was exhumed with all its bedform characteristics.

The mid-channel bars that occur farthest from the seaface are generally muddy, whereas, those nearer the sea are mostly sandy. Some mid-channel bars that occur

Table 3.2 Length, breadth and length-breadth ratio (L/B) of the mid channel bars of the Thakuran river

Locality	Length (L) (in km)	Breadth (B) (in km)	L/B
Bhubaneswari Dwip	3.15	1.05	3.0
Damkal Dwip	2.1	1.0	2.1
Paschim Sripatinagar	3.5	3.7	0.95
Upendranagar	2.5	1.1	2.27
Rakhalpur	2.45	0.9	2.72
Sridharnagar	3.7	2.8	1.32
Dhanchi I	3.0	1.25	2.4
Dhanch II	4.75	1.1	4.32

in the middle stretches of the rivers exhibit excellent grain-size differentiation from their northern (landward) to the southern (seaward) extremities. Generally the southern halves of these bars are sandy and the northern halves are muddy in nature. In the northern half of Paschim Sripatinagar mid-channel bar, sporadic occurrences of small-scale longitudinal ripples and wave ripples are observed. Megaripples are generally absent in the mudflats. The surface sediments are dominantly muddy with clayey silt. Graphic mean size (M_Z) of surface sediments ranges from 4.86 to 6.45 phi and sorting from 0.47 to 1.45 phi, i.e. well sorted to moderately sorted (Das 2016a). The southern portion of the sandy mid-channel bar is sculptured with ripples, megaripples and sand waves of various morphological types and scales. The area, thus, displays relief features of different orders on the bar surface. The surface sediments have mean grain size (M_Z) from 2.02 to 3.55 phi (medium to fine sand), sorting (σ_1) from 0.22 to 0.84 phi (very well sorted to moderately well sorted). The influence of flood flow is reflected in the trains of sand waves and megaripples. The ebb flow, however, besides changing the morphology of the flood-oriented bedforms, is also responsible for the generation of small-scale ripples. The flood and ebb flow directions over this mid-channel bar have been interpreted from the orientation of various bedforms (Das 2016b).

A large number of meandering creeks traverses this mid-channel bar exposing its lamination patterns. The creeks are of variable length (20–500 m) and width (up to 15 m). Lee-slope orientations of current ripples generated along these creeks are often different from those occurring outside the creek boundaries. Slumping and collapsing of creek margins and concurrent deposition of the eroded material at the mouth of the creeks develop both sandy and muddy micro-deltas (Fig. 3.7).

Textural studies of samples collected up to a depth of 120 cm with the help of an auger revealed a general coarsening upward of mean grain size (M_Z), i.e. a 'Cu' sequence (Fig. 3.8). This upward coarsening of mean grain size is, however, noticed in the mid-channel-bar sediments occurring up to a 40 km stretch upstream from the mouth of the Thakuran River. This is perhaps due to the landward advancement of the river-mouth sand caused by strong waves and flood currents.

3.3.1.7 Swash Platform

The crescentic bar seen near the mouth of the Thakuran River at Dhanchi experiences wave rush-up and backwash in its intertidal zone. This zone behaves like swash platforms (Oertel 1972) where the accretionary sandy bar sediments are reworked by wave action and exhibit features like that of sandy beaches. Instead of seaward dip, the inclination of the swash-bar surface is towards the funnelled estuary. The elongation axis of the swash platforms is almost at right angles to the shoreline (Fig. 3.9).

Morphologically, the swash platforms show gently dipping surfaces (4°–5°) with minor undulation on the higher topographic areas. These undulations are due to the

Fig. 3.7 A muddy microdelta on the slope of a creek. Note the transverse growth marks on the surface and lobed nature of the lee-side

presence of the backwash ripples or low amplitude antidunes which characterise the upper foreshore zone of most beaches. The maximum advance of wave swash is marked by swash marks (Fig. 3.10) made up of wood pieces, weeds and mangrove leaves. The mid-intertidal zone of the platforms shows major undulations due to the presence of megaripple trains. Some convex upward ridges on the platforms are ornamented both by large-scale and small-scale rhombhoid marks. The lowermost stretch of the platforms is characterised by abundant current crescents, rill marks and small-scale wave ripples. At the distal end of the platforms where they merge into the muddy mangrove banks, the sandy swash platforms are underlain by cohesive mud that supports luxuriant mangrove vegetation, particularly *Phoenix paludosa* forests, the natural abode for the famous royal Bengal tiger of Sunderbans. The overlying mud encroaches on the sandy platform to show a sinuous train. Sometimes, portions of upper muddy bank sink down to the low-lying sandy substratum resulting in entrapment of mud chunks within sandy units. Thus the surface roughness of the swash platform is the manifestation of bedform patterns and their spacing characteristics.

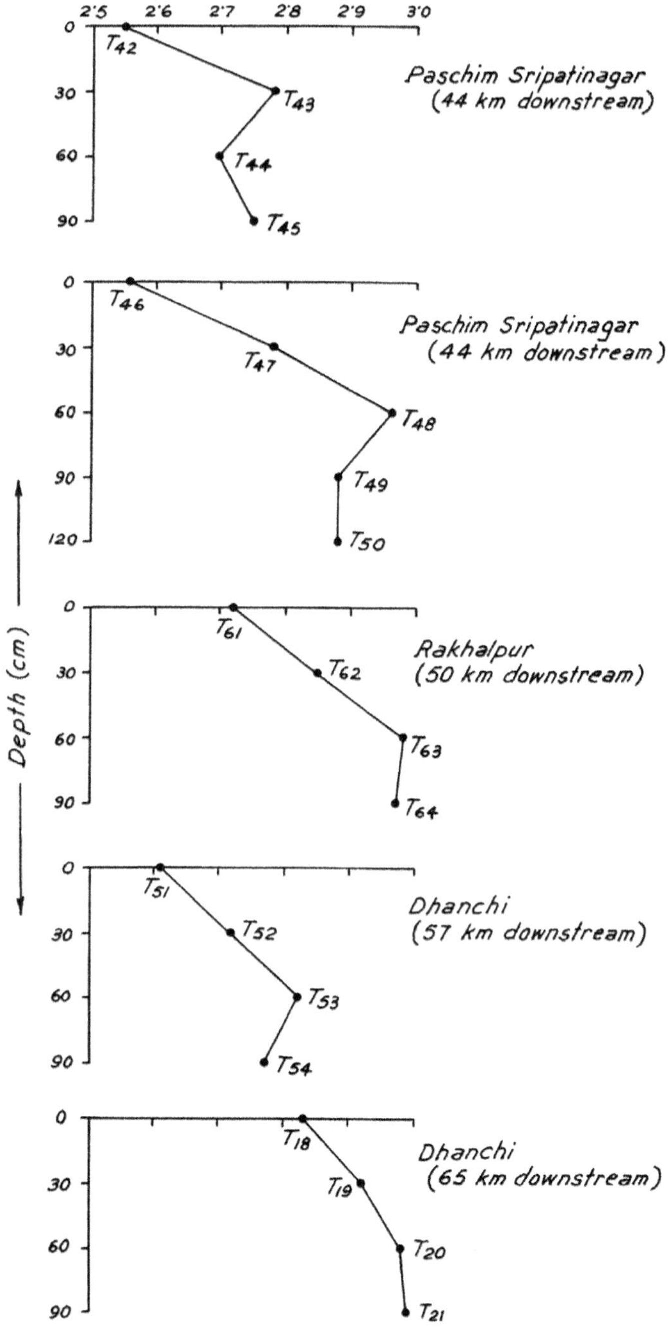

Fig. 3.8 Coarsening upward (CU) sequence in mid channel bar samples of Thakuran River

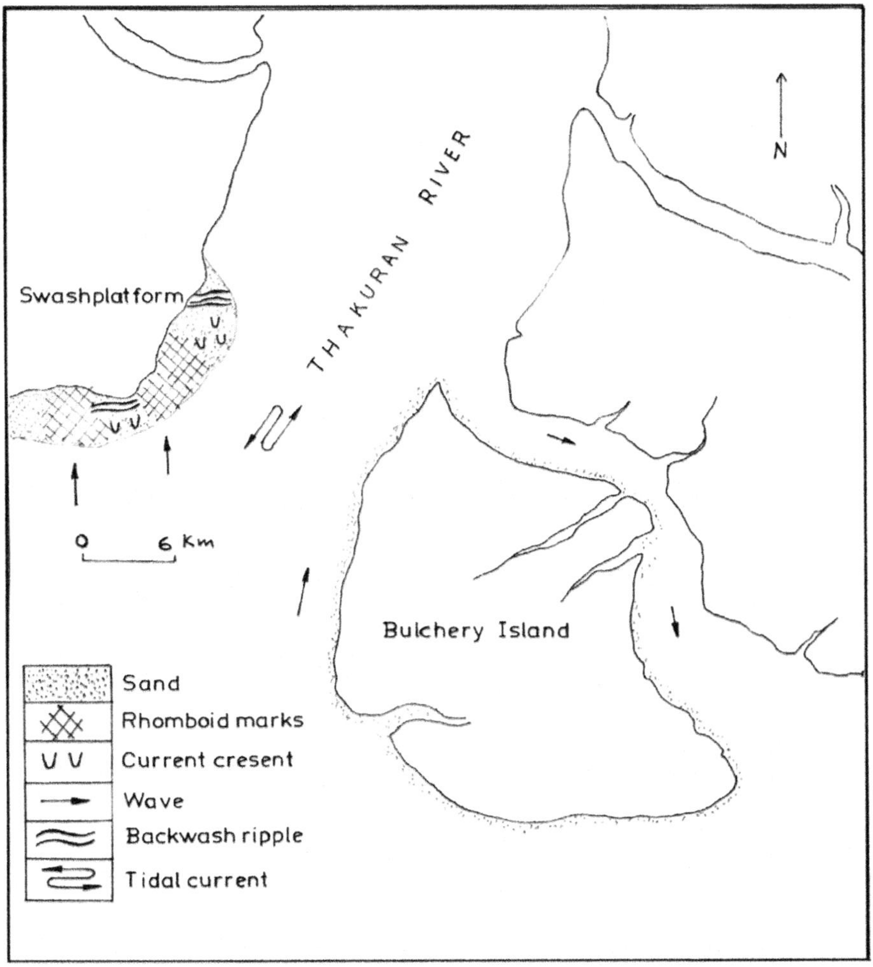

Fig. 3.9 Map showing location of swash platform. Grain size and surface sedimentary structures are also shown

3.3.1.8 Wash-Over Flats

The wash-over flats occupies a very small part of the geomorphological spectrum of the Thakuran River basin. South-west of the Dhanchi Island at the mouth of river, a wide sandy wash—over flat of channel-fill origin has been identified. The flat has a gradual seaward slope ranging from 3° to 6°.

The wash over flat is generally intertidal with ripples and rills as the dominating surface features. The rills, produced mainly by the backwash flow, make a complex network of branching up to the seaface. The upper supratidal portion is characterised by a series of aeolian dunes, many of which are conspicuous by their

Fig. 3.10 Swash marks on sandy swash platform. Convexity of the marks is in the riverward direction. A current crescent is seen to the *right* of the photograph. Opening of the arms of the current crescent also indicates the same riverward direction. The gastropod shell has been drifted up by waves after the formation of current crescent. Knife is 26 cm

accumulation of mangrove leaves. Exotic boulders and variably-sized oyster shells are strewn all over the flat and give definite evidence of the repeated reworking of the surficial sediments. The stability of the flats is greatly reduced by the felling of mangrove trees, the upright trunks of which stand bare on the sandy wash-over flat surface.

3.3.1.9 Ebb-Tidal Delta

Bulchery Island exhibits the most prominent sites of sedimentation at the confluence of the Thakuran River and the open sea. This huge island (approximately 7 km across) is an ebb-tidal delta. The narrow Gokultali creek (locally known as Gokultali gang) is the ebb channel whereas, the course of Thakuran River acts as the flood channel. The upper surface of this island remains exposed even at high tides throughout the year. The island is truly an ebb-tidal delta (Hayes 1976). The northern part of the mouth bar is protected from wave action and is under the influence of tidal currents. The southern part, on the other hand, faces the Bay of Bengal and experiences severe wave attack. The spring tidal amplitude is as high as 6 m or more (macrotidal). This ebb-tidal delta or the river-mouth bar forms where the ebb-flow is countered by the incoming refracted waves from the Bay of Bengal (Hubbard et al. 1979). According to Hayes (1976), the ebb-tidal deltas are more characteristics of a mesotidal setting. Elliot and Gardiner (1981), however, reported their occurrence in a macrotidal framework.

The intertidal and supratidal deposits of Bulchery Island are clearly discernible. The intertidal zone exhibits dovetailing of sand and mud for an expanse of 800 m. The supratidal zone extends 170–180 m and is composed of non-cohesive sands of reddish- yellow colour implying a greater rate of oxidation. Bioturbational effects are quite common. A gradual change from intertidal to supratidal environment is readily recognisable from the marginal to the topmost part of the island. The supratidal sands are very fine and well-sorted. The island is dissected by a number of tidal creeks of sinuous nature. These creeks are generally ephemeral and are expected to change their courses with time. Natural creek sections often reveal alternations of sand and mud layers of variable thickness ranging between 10 and 45 cm.

3.4 Sand-Body Geometry of the Geomorphic Zone

Four different types of sand bodies characteristic of four geomorphic regimes have been recognised in the Thakuran basin (Table 3.3). The dimensions of sand bodies, grain size, bedforms, and internal sedimentary structures are the principal parameters that characterise each of them (Coleman et al. 1988; Harris 1988; Terwindt 1988; Banerjee 1989). Despite some overlapping, these parameters clearly indicate the different types of recognisable bodies of sand.

Table 3.3 Diagenostic features of four basic types of sand bodies of different geomorphic zones

Parameters of sand body geometry	Mid-channel bar	River bank (natural levees)	Point bar/Swash Platform	Ebb Tidal Delta
Dimension	4 km × 3.54 km × 14 m	6 km × 1.5 km × 18 m	5 km × 0.5 km × 17 m	9 km × 6 km × 16 m
Grain size (Phi)	2.02–2.96	2.62–2.92	2.58–2.61	2.53–2.77
Bedforms	Sandwaves, Megaripples, Ripples, (linguoid, ladder-back, flat-topped), Rill marks	Small-scale ripples	Megaripples, Reversing ripples, Backwash ripples, Swash marks, Current crescents, Rill marks	Sandwaves, Megaripples, Ripples
Internal structure	Large-scale cross-bedding, Alternation of sand and mud couplets, Parallel stratification	Ripple-drift lamination, Parallel lamination	Epsilon cross-bedding, Alternate sand and mud lamination	Large and small-scale cross bedding
Facies Sequence	Coarsening upward	Fining upward	Fining upward	Coarsening upward

3.5 Summary

The geomorphic environments and facies of the Thakuran River basin are diverse and intermingle in both the lateral and vertical dimensions. The subaerial activities are often mixed with subaqueous processes. The shallow subaqueous deposits are again transported to the deeper sea by submarine canyons. Recycling of shallow coastal deposits often disturbs the sediment-distribution patterns and imports temporal and spatial variations into the sedimentary budget.

The coastal zone at the confluence of the Thakuran River and the Bay of Bengal reveals various apparently accretional, erosional sectors without showing a significant progradation of the shoreline over time. This is because the bulk of the sedimentary load transported from the upland contributes to the Bengal deep-sea fan—a mechanism similar to that described in the Bruun Rule that states that a generally sandy coastline with an equilibrium profile will retreat (or remains steady!) in response to a rise of sea level relative to the land, until the profile has been restored by seafloor accretion.

References

Allen JRL (1963) The classification of cross-stratified units with notes on their origin. Sedimentology 2:93–114

Banerjee I (1989) Tidal structures in the Glauconitic Sandstone, Countess Field, Southern Alberta, Canada. In: Reinson GE (ed) Modern and ancient examples of clastic tidal deposits—a core and peel workshop. Canadian Society of Petroleum Geologists, Second International Research Symposium on Clastic Tidal Deposits, Aug 22–25. Calgary, Alberta, pp 89–97 (126p)

Bhattacharya A (1999) Embankments and their ecological impacts: a case study from the Tropical low-lying coastal plains of the deltaic Sunderbans, India. In: Large—scale constructions in coastal environments. Springer, Berlin, pp 171–180 (194p)

Bridges PH, Leeder MR (1976) Sedimentary model for intertidal mudflats channel with examples from the Solway Firth, Scotland. Sedimentology 23:533–552

Coleman SM, Berquist CR Jr, Hobbs CH III (1988) Structure, age and origin of the bay mouth shoal deposits. Chesapeake Bay, Virginia. Mar Geol 83:95–113

Das GK (2015) Estuarine morphodynamics of the Sunderbans. Coastal research library, vol 11. Springer, Switzerland, 211p

Das GK (2016a) Sediment grain size. In: Kennish MJ (ed) Encyclopedia of Estuaries. Springer, Berlin, pp 555–558 (760p)

Das GK (2016b) Sedimentary structures. In: Kennish MJ (ed) Encyclopedia of Estuaries. Springer, Berlin, pp 568–572 (760p)

Elliot T, Gardiner AR (1981) Ripple, megaripple and sandwave bedforms in the macrotidal Loughor Estuary, South Wales, UK. Spl Publ Int Asso Sed 5:51–64

Harris PR (1988) Large-scale bedforms as indicators of mutually evasive sand transport and sequential infilling of wide-mouthed estuaries. Sed Geol 57:273–298

Hayes MO (1976) Morphology of sand accumulation in estuaries: an introduction to the symposium. In: Cromn LE (ed) Estuarine research, vol 2, Geology and engineering, Academic Press. New York, pp 3–22

Hubbard DK, Oertal G, Nummedal D (1979) The role of waves and tidal currents in the development of tidal inlet sedimentary structures and sand body geometry: examples from North Carolina, South Carolina and Georgia. J Sed Petrol 49:1073–1092

Mossop GD, Flach PD (1983) Deep channel sedimentation in lower Cretaceous McMurry Formation, Athabasca oil sands, Alberta. Sedimentology 30:393–409

Oertel GE (1972) Sediment transport on estuary entrance shoals and the formation of swash platform. J Sed Petrol 42(4):528–563

Oomkens E, Terwindt JHJ (1960) Inshore estuarine sediments in the Haringvliet (Netherlands). Geol Mijnbouw 39:701–710

Owens R (1981) Holocene sedimentation in the north-western North Sea. In: Shuttenhelm RJE, Van Weering Tj CE (eds) Holocene marine sedimentation in the North Sea basin. SD Nio, Spec Publs Int Asso Sediment, pp 303–322 (515p)

Pantin HM (1969) The appearance and origin of colours in muddy marine sediments around New Zealand. N Z J Geol Geophys 12:51–66

Rahmani RA (1988) Estuarine tidal channel and nearshore sedimentation of a Late Cretaceous epicontinental sea, Drumheller, Alberta, Canada. In: deBoer PL, VanGelder A, Nio SD, (eds) Tide-influenced sedimentary environments and facies. D Reidel Publishing Company, pp 433–471 (530p)

Reineck HE, Singh IB (1980) Depositional sedimentary environments. Springer, New York 408p

Sarkar SK, Basumallick S (1968) Morphology, structure and evolution of a channel island in the Barakar River, Barakar, West Bengal. J Sed Petrol 38:747–754

Smith DG (1988) Modern point bar deposits analogous to the Athabasca oil sands, Alberta, Canada. In: deBoer PL, VanGelder A, Nio SD (eds) Tide-influenced sedimentary environments and facies. D Reidel Publishing Company, pp 417–432 (530p)

Stanley DJ, Hait AK (2000) Holocene depositional patterns, neotectonics and Sunderbans mangroves in the Western Ganges—Brahmaputra delta. J Coast Res 16(1):26–39

Swift DJP, Boehmer WR (1972) Brown and grey sands on the Virginia Shelf: colour as a function of grain size. Bull Geol Soc Am, 877–884

Terwindt JHJ (1988) Paleo-tidal reconstruction of inshore tidal depositional environments. In: deBoer PL, VanGelder A, Nio SD (eds) Tide-influenced sedimentary environments and facies. D Reidel Publishing Company, pp 233–264 (530p)

Van Straaten LMJU (1961) Sedimentation in tidal flats areas. J Alberta Soc Petrol Geol 9:203–213

Chapter 4
Sediment Composition

Abstract Thakuran River sediments are typically estuarine. Lithogenic constituents are dominant with about 85–90 % in bulk and biogenic components constitute the rest (10–15 %). Quartz, feldspars, mica, lithic fragments and some heavy minerals are the main terrigenous constituents. Secondary overgrowth of quartz with multiple rims of inclusions suggests their derivation from sedimentary rocks of different cycles of sedimentation. The heavy mineral assemblage in the sediments leads to the conclusion that the sediments have been derived from the acidic igneous rocks of the Himalayas in the north and the metamorphic rocks of the Precambrian shields of the west and north-west of Bengal Basin. The benthic foraminiferal assemblage indicates a shallow-water, moderate-to-low saline environment.

Keywords Lithogenic components · Biogenic components · Heavy minerals · Acid igneous minerals · Metamorphic sources · Sedimentary rocks · Thakuran river · Sunderbans

The Thakuran drainage basin of the Sunderbans is characterised by the low-lying alluvial tropical coastal plains of the Ganges-Brahmaputra delta. The sediments of the many distinctive geomorphic zones like point bar, mid-channel bar, marginal bar, natural levee etc. are siliciclastic with sand-silt-clay as the chief constituent. The sand is characteristically terrigenous in nature and mainly composed of quartz with pockets of heavy mineral concentration (Das 2015). Quartz grains from the sediment deeper than 0.5 m in trench sections often show yellowish-brown surficial tint. Biogenic and terrigenous or detrital components constitute the common coarser fraction of the sandy materials. Quartz, feldspar, muscovite, biotite etc. constitute the terrigenous materials, contributing about 40–80 % of the total coarse fraction. Total heavy mineral concentration combining biotite, muscovite, amphiboles, pyroxenes, epidote-zoisites is moderately high. Heavy minerals like rutile,

G.K. Das, *Tidal Sedimentation of the Sunderban's Thakuran Basin*,
DOI 10.1007/978-3-319-44191-7_4

monazite, zircon, garnet, opaques etc. are considerably less concentrated in all the collected samples. Biogenic materials are very few in the clayey silt samples, whereas these are abundant in the sand and silty-sand types of sediments. The non-detrital fraction includes skeletal grains of gastropods and bivalves or their fragments and calcareous concretions of possible algal origin. Benthic foraminifers with some pteropods, crinoidal stems and echinoid spines occur only as a very small part in the coarser fraction of the sandy materials.

A systematic study of both lithogenic and biogenic content of the sediments of the Thakuran River might possibly help for assessing the impact of the estuarine environment. Laboratory work concerning sediment properties has consisted of mineralogical and granulometric analyses of the collected samples. Mineralogical studies of both heavy and light fractions were done. The heavy minerals were separated from the lights by the heavy-liquid (bromoform) separation method (Griffiths 1967). Grain-slides and impregnated sediment sections were studied under the petrological and high-power binocular microscope to identify the minerals and their textural properties.

4.1 Lithogenic and Biogenic Components

The siliciclastic sandy, silty sand and clayey silt sediments of the Sunderbans show some mineralogical variations from the landward to the seaward stretches of the river, as well as from one geomorphic zone to another. Primarily, both lithogenic (85–90 %) and biogenic (10–15 %) components constitute the mineralogical composition of the sediments.

Among the lithogenic components, light mineral fractions constitute between 96 and 99 % of the sediments, whereas, the heavy minerals (specific gravity >2.9) constitute only 0.5–3.7 %. The light mineral fraction is composed chiefly of quartz, feldspars of microcline, orthoclase and sodic plagioclase varieties, muscovite and lithic fragments of schistose, semi-schistose (phyllitic), cherty and quartzitic rocks (Fig. 4.1). The heavy minerals, in decreasing order of abundance, are magnetite, garnet of white and pink varieties, hornblende, epidote-zoisite, biotite, chlorite, kyanite, sphene and zircon (Fig. 4.2).

The biogenic fraction of sediments is composed of both animal and plant bodies. Fragmentary macrobenthic and meiobenthic fauna as well as the entire shells of microbenthic fauna (Table 4.1) are important among the biogenic fractions. All these shells are composed of $CaCO_3$ and are susceptible to bioerosion. Entire shells of microbenthic bivalves, gastropods and foraminifers are quite common. The fragmentary animal shells are chiefly of macrobenthic and meiobenthic bivalves, gastropods and crustaceans. The assemblage of foraminifers like *Elphidium* sp, *Ammonia* sp, *Asterorotalian pulchella*, *Triloculina* sp and *Parafissurina* sp indicates a shallow-water, moderate–to-low saline environment (Jansen and Hensey 1981).

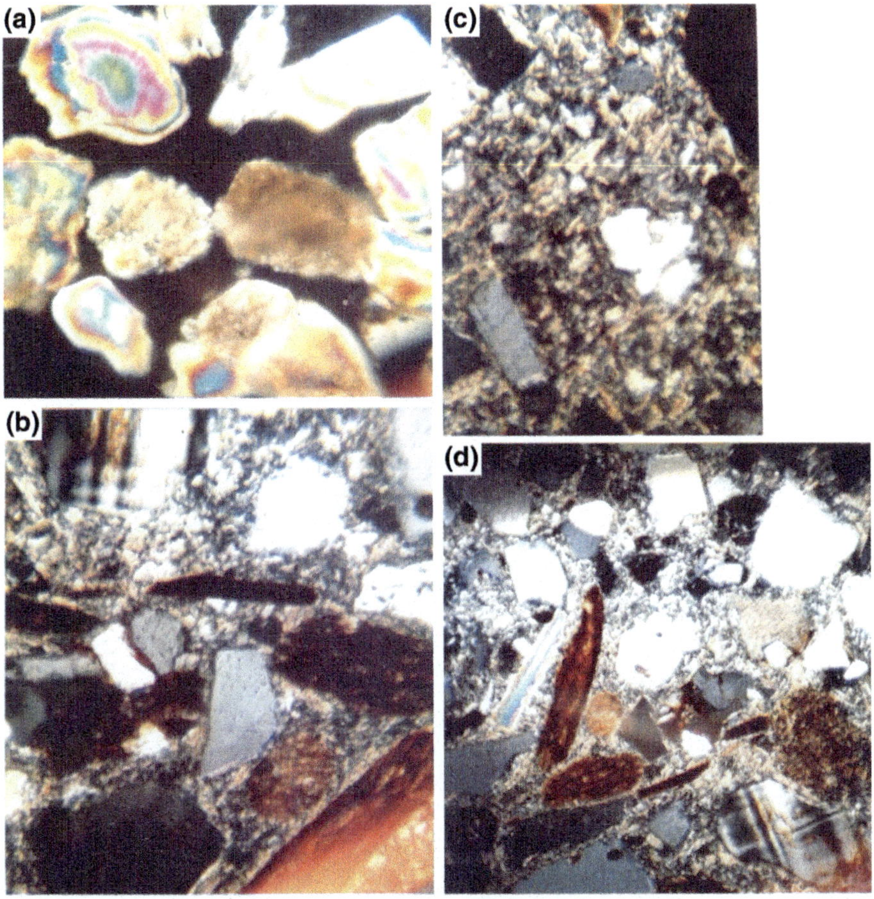

Fig. 4.1 **a** Loose grains mounted in glass slides. Grains of quartz, feldspar and rock fragments show, colour zoning because of greater thickness. Magnification: X 10, **b** Thin section of impregnated grains between crossed nicols. Grains of microcline (crosshatched), orthoclase, quartz, schistose rock fragment, biotite (*dark brown*) and opaques are seen. Magnification: X 10, **c** Thin section of impregnated grains. Composites quartz grains, orthoclase (*grey* coloured) opaques are present. Magnification: X 10, **d** Thin section of impregnated grains between crossed nicols. Grains of quartz, rock fragments, biotite (*elongated brown*) muscovite (*elongated bluish green*), orthoclase are seen. Note the subangular to subround nature of grains. Magnification: X 10

Fresh and decomposed fragments of mangrove trunks, leaves, roots and pneumatophores are inevitably present in the sediments in variable proportions. This plant debris is generally greater in sediments in proximity to the mangrove forests. Thus the riverbank and point-bar samples contain a great abundance of plant debris. This is well evidenced in trench sections where huge mangrove woods constitute a

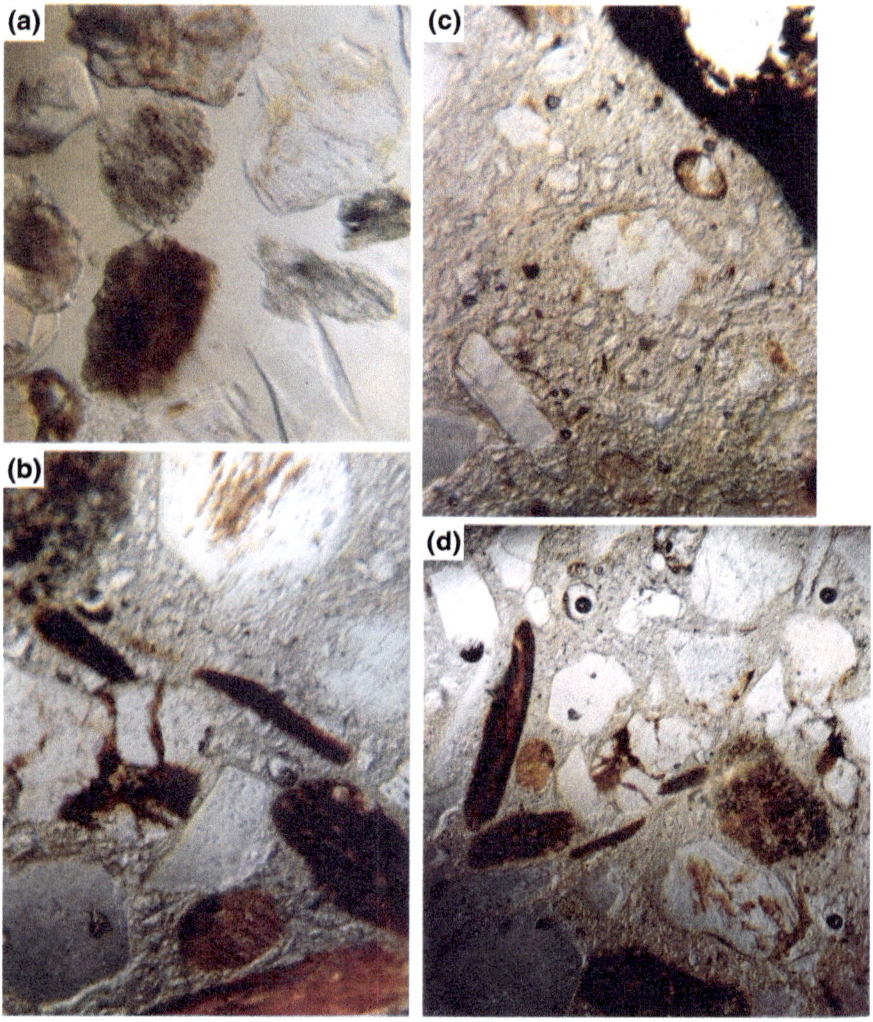

Fig. 4.2 **a** Loose grains mounted in glass slide. Grains of feldspar, epidote-zoisite, chlorite, sphene and kyanite are present. Magnification: X 10, **b** Thin section of impregnated grains between crossed nicols. Section shows euhedral feldspar, quartz and biotite. Magnification: X 10, **c** Thin section of impregnated grains. Section shows several grains of orthoclase, few quartz, biotite and sphene. Magnification: X 10, **d** Thin section of impregnated grains. Grains of orthoclase, quartz, biotite, chlorite and hornblende are present. Magnification: X 10

major component of large-scale planar tabular cross-beds. The deeper samples of mudflats associated with point bars and mid-channel bars also reveal large amounts of humus material because of the decomposition of mangrove litters.

Table 4.1 Components of sediment composition of the estuarine river sediments of the Sunderbans and the probable source rocks

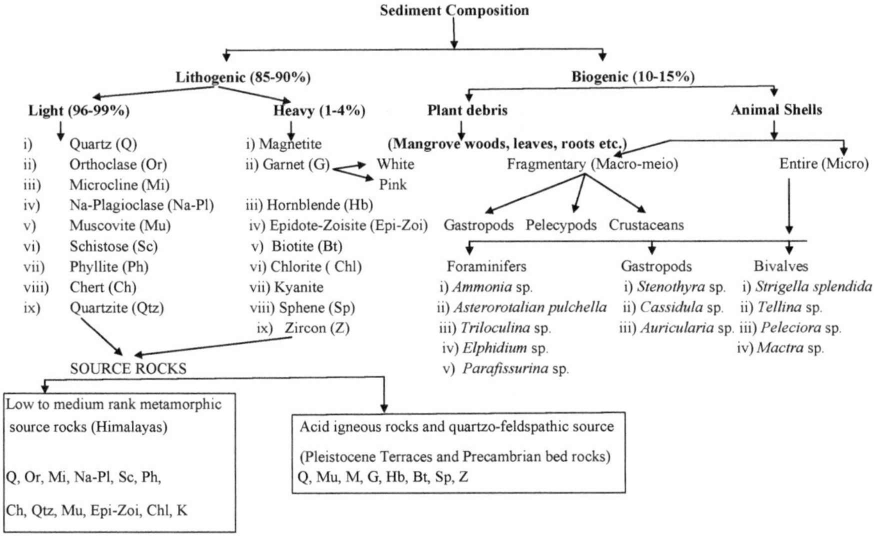

4.2 Source Rock of River Sediments

The sediments of the Thakuran River are primarily derived from the sedimentary and metasedimentary rocks of the Himalayas in the north down through the primary drainage systems of the Ganges and the Brahmaputra rivers. Furthermore, the sediments fed by the Pleistocene terraces flanking the delta on all sides and the acidic-igneous bedrocks of the Precambrian shields of the north-west and west have a significant role in the delta formation.

The association of light and heavy minerals, namely quartz, sodic plagioclase, muscovite, biotite, garnet, hornblende, kyanite, epidote-zoisite, magnetite, phyllitic and quartzite rock fragments indicates low-to-medium rank metamorphic source rocks of the Himalayas (Table 4.1). The presence of quartz, K-feldspers as microcline and orthoclase, magnetite, muscovite, biotite, hornblende, sphene and zircon (Figs. 4.1 and 4.2) in the sediments refers to an acidic-igneous source (Pettijohn 1984) related to the Pleistocene terraces and the Precambrian shields (Table 4.1). The detrital quartz occurs both as single and composite grains (poly-crystalline quartz). Secondary overgrowth in quartz grains with two to three rims of inclusions suggests their derivation from sedimentary rocks of twotothree cycles of sedimentation.

4.3 Summary

Mineralogically, both lithogenic (85–90 %) and biogenic (10–15 %) components constitute the sediments of the different geomorphic bodies of the Thakuran River. The light fractions are chiefly composed of quartz, feldspar of various types, and a few rock fragments. The detrital quartz occurs both as single and composite grains (polycrystalline quartz). Secondary overgrowth in quartz grains with two or three rims of inclusions suggest their derivation from sedimentary rocks of two to three cycles of sedimentation. Heavy minerals constitute 0.5–3.7 %. The heavy mineral content refers to both acidic igneous and metamorphic sources. The acidic igneous minerals are derived chiefly from the Himalayas in the north, and the metamorphic mineral assemblage speaks of their derivation from the Precambrian shields of the west and northwest of the Bengal basin.

Mineral suits including both light and heavy fractions have been examined to infer the acidic igneous and metamorphic provenance. The assemblage of heavy minerals refers to a major Himalayan component of the source derivation of the sediments, together with contributions from the Precambrian terrains and Pleistocene terraces.

Mechanical fragmentation of both gastropods and bivalves indicates a nearshore coastal marine environment with high-energy conditions. Presence of benthic foraminiferal assemblages with comparatively smaller sizes suggest a euryhaline a shallow-marine environment as well as a huge influx of freshwater from the Hooghly estuary in the eastern boundary of the Sunderbans. The higher population density and species diversity of the foraminifers indicate a relatively slow rate of sedimentation.

References

Das GK (2015) Estuarine morphodynamics of the Sunderbans. Coastal research library, vol 11. Springer, Switzerland, 211p

Griffiths JC (1967) Scientific method in analysis of sediments. McGraw Hill Book Company, 508p

Jansen JHJ, Hensey AM (1981) Interglacial and Holocene sedimentation in the northern North Sea: an example of Femian deposite in the Tartan Field. Spl Publ Inst Asso Sediment 5: 323–334

Pettijohn FJ (1984) Sedimentary rocks. 3rd edn. CBS Publishers & Distributors. India, 628p

Chapter 5
Mudflats and Tidal Creek Sedimentation

Abstract The Thakuran River is truly a major tidal creek in the low-lying, tropical coastal plains of the Ganges-Brahmaputra delta. It is fed by to-and-fro-moving flood and ebb flows without any headwater discharge. The Thakuran River basin is intersected by numerous creeks and characterised by the mudflats on the river margins as well as in the riverbed. Lateral sedimentation takes place on meandering tidal creeks and gullies, whereas vertical sedimentation is found in the mudflats. Morphological features like mud ridges, mud microdelta, mud pellets, and mud lumps are recognised in their relationships to creek sedimentation. Depositional features typical of tidal environment like mud couplets, tidal bedding and tidal bundles have been recognised. The laminae of sand and clay in the core samples from the mudflat indicate a gradual decrease in current velocity in the upward direction.

Keywords Tidal creek · Mudflats · Mud ridges · Mud microdelta · Mud pellets · Mud lumps · Wood clumps · Convolute lamination · Thakuran river · Sunderbans

Mudflats predominate in the low-energy areas of suspensional depositions, particularly in the middle to the upper stretches of the rivers belonging to meso-microtidal settings. The areas of mud deposition include certain portions of mid-channel bars, point-bars and margins of tidal creeks that often intersect the intertidal flats. Mudflats occur, covering a few square metres to hundreds of square metre areas, with the thickness of the mud blankets ranging from a few centimetres to less than a metre. Fine rhythmic laminations of silt and clay characterise the deposits. Lenses of silty sand of 2.12 mm thickness occasionally intercalate the mud deposits. Physical and rheological properties (viscosity, plasticity and thixotropy) are known to be governing factors for erosion and resuspension of fine-grained sediments in tidal river and estuarine environments (Faas 1981; Das 2015). Generally the tidal mud deposits are soft, grey and water saturated. The water content varies at different levels of the accumulated mud.

G.K. Das, *Tidal Sedimentation of the Sunderban's Thakuran Basin*,
DOI 10.1007/978-3-319-44191-7_5

5.1 Physical Parameters of Muddy Sediments

Generally the tidal mud deposits are soft, grey and water-saturated. The water content varies at different levels of the accumulated mud. The lower parts of mud deposits are generally more plastic and exhibit greater penetrability compared to that of the upper portions (Table 5.1). The water content in some areas is sometimes so much that, at times of field work, people may sink up to their knees.

Granulometric properties reveal a range of silt content varying from 49.40 to 99.57 %, clay content from 0.14 to 34.41 % and mean size from 4.45 to 7.7 phi (Table 5.2). Organic-matter content is always higher in the upper mudflats with mangrove roots and pneumatophores than that of lower mudflats. Mineralogically, the silty fraction is composed of 85.90 % quartz, 5.6 % feldspar, and the rest mica. Generally, micas concentrate more towards the upper and landward stretch of the river with low tidal energy conditions. The dominance of quartz (with less feldspar) characterise the siliciclastic nature of the mudflats. Micas, because of their platy or flaky habit, are very sensitive to even slight changes in the energy conditions. Their absence or minimal presence in many occasions in the high-energy seaward stretch

Table 5.1 Relative penetrability in the mudflat sediments of the Damkal Dwip

Profile no. (Spacing of profiles 6 m)	Penetration depth (cm) in mudflat (Average of 4 readings)		
	Upper mudflat	*Middle mudflat*	*Lower mudflat*
I	6.6	7.2	12.9
II	9.7	10.2	14.3
III	2.9	3.6	5.1
IV	5.2	8.6	8.9
V	6.7	8.2	9.1
VI	9.2	9.4	11.6
VII	7.8	10.8	10.9
VIII	6.7	9.5	11.2
IX	8.4	12.6	12.1
X	10.2	11.7	11.9
XI	9.3	14.7	15.1
XII	6.4	7.5	9.1
XIII	6.5	8.4	12.6
XIV	2.8	6.8	7.2
XV	3.8	7.8	8.6
XVI	4.5	5.6	8.2
XVII	5.7	5.9	7.9
XVIII	9.3	11.4	12.6

Table 5.2 Statistical size parameters and sand-silt-clay percentage of the mudflat samples of the Thakuran River

Sample Locations (Distance in km from landward to seaward direction)	Sample no	Graphic Mean size (Mz)	Inclusive Graphic Standard Deviation (σ_1)	Inclusive Graphic Skewness (SK_1)	Graphic Kurtosis (K_G)	Sand %	Silt %	Clay %
Bhubankhali (10 km)	T_{70}	7.23	1.15	−0.27	0.92	3.20	96.80	–
Chuprijhara (14–16 km)	T_{67}	6.40	0.52	−0.17	1.76	1.10	98.90	–
	T_{68}	7.70	1.86	0.12	0.66	0.89	63.70	35.41
	T_{69}	6.65	0.62	−0.29	1.37	0.43	99.57	–
Bhubaneswari Dwip (25 km)	B1	6.0	0.99	−0.05	1.47	5.04	91.02	3.94
	B2	6.43	0.66	−0.07	2.19	4.42	94.94	0.64
	B3	6.31	0.71	−0.09	1.39	4.33	90.76	4.91
	B4	6.39	0.69	−0.13	1.98	5.34	91.39	3.27
Paschim Sripatinagar (41 km)	T1	5.95	0.73	−0.39	0.96	11.74	87.88	0.38
	T2	5.90	0.84	−0.30	0.66	22.61	72.88	4.51
	T3	5.96	0.96	−0.15	0.93	13.14	85.42	1.44
	T4	5.77	0.63	−0.43	0.74	12.94	86.92	0.14
	T5	6.30	1.08	−0.30	0.69	18.49	77.73	3.78
Upendranagar (49 km)	L_1	6.31	1.34	−0.69	0.70	6.44	93.20	0.36
	L_2	5.58	1.69	−0.66	0.64	24.24	72.12	3.64
	L_3	4.45	1.80	0.32	0.60	49.84	49.40	0.76
	L_4	2.58	0.32	−0.34	1.50	96.61	3.39	–
	L_5	4.66	1.64	0.32	0.68	49.48	50.52	–
	L_6	2.61	0.21	−0.22	1.22	100.0	–	–

is attributed to winnowing or bypassing effects. These flaky minerals sometimes behave as the hydraulic equivalent to silt, and winnowing processes effectively transport them from the middle to the upper stretches of the lower-energy environment (Van Weering 1981). The distribution of micas thus indicates a sedimentation mechanism due to gradual decrease in energy towards the landward direction. This is further reflected by the distribution pattern of clear sand towards the seaface of the river and more silt, clay and micas progressively towards the landward direction. Biogenic components of the mud deposits rarely exceed 10–15 % and are composed of fine mangrove debris, fragmentary and entire shells of crabs, gastropods, bivalves and foraminifers (Table 5.3). Fecal materials of these animals also constitute a part of the sediments.

Table 5.3 A list of crabs, gastropods, bivalves and foraminifers from the mudflat sediments of the Thakuran River

CRABS	FORAMINIFERS
Scylla serrata	*Ammonia* sp
Macrophthalmus sp	*Asterorotalian pulchella*
Metaplax crenulata	*Triloculina* sp
Ocypode macrocera	*Elphidium* sp
Uca acuta acuta	*Parafissurina* sp
Illyoplex gangetica	
GASTROPODS	BIVALVES
Stenothyra deltae	*Mactra luzonica*
Cassidula nucleus	*Tellina corlroloides*
Ellobium gangetica	*Strigilla splendid*
Telescopium telescopium	*Pelecyora sp*
Cerithidea cingulata	*Anadara granosa*
Littorina melanostoma	*Singuilaria accuminata*
Nerita articulate	*Macoma birmanica*
Nassarius fovelatus	*Crassostrea cuculata*
Cymia carniferea	*Meretrix meretrix*
Onchidium tigrinum	*Placenta placenta*
Assiminea brevicula	*Teredo sp*
	Paphia malabarica

5.2 Sedimentation Types on Mudflats

Mudflat sedimentations are of two main types: (i) lateral sedimentation and (ii) vertical sedimentation. Lateral sedimentation takes place on meandering tidal creeks and gullies and results in dipping laminae. Deposition here is rather rapid. Vertical sedimentation, on the other hand, is found on the higher parts of the mudflats, above the influence of tidal creeks. Here sedimentation is rather slow and yields horizontal or subhorizontal laminae. Both inclined and horizontal mud laminae intermingle near edges of the raised mudflats.

5.3 Sedimentation on a Tidal Creek

Intertidal mudflats on point bars bordering the Thakuran River are very commonly dissected by tidal creeks. The creeks are 10–25 m long and 2–10 m wide. They are sinuous to meandering in pattern and act as pathways for both flood and ebb flows depending on alternation of tidal currents (Fig. 5.1). On many occasions, interlocking ridges and spurs along meandering paths characterise the creek morphology. The mud deposits bordering creeks are well stratified, but the strata may be obscured because of local slumping. They are slightly inclined near creek margins but become horizontal to subhorizontal on the upper mudflats. The creek bottoms

Fig. 5.1 A meandering tidal creek on the mudflat of Thakuran River. Note the interlocking ridges and spurs

are generally composed of silty sand, which generates small ripples along the creek axis. Various morphological features are recognisable in creek sedimentation.

5.3.1 *Mud Ridges*

Small mud ridges normal to creek axes develop on steeply inclined (40° to −5°) creek slopes. Freshly deposited mud on creek slopes suffers differential erosion because of the drainage of water through intermittent rills. The non-eroded portions remain as mud ridges on creek slopes. The ridges are 12–15 cm broad, 5–10 cm high and 40–45 cm long. They usually occur in complex groups and show less bifurcation compared to the types described by Reineck and Singh (1980). Many of these mud ridges are also produced by liquefaction of mud already settled along creek slopes. Reineck and Singh (1980) described similar mud ridges from ancient tidal flats of Devonian age.

5.3.2 *Mud Microdelta*

Microdeltas are formed by the accumulation of mud on the creek floor. These are triangular in plan and wedge-shaped in longitudinal section. Their surface is slightly convex upward. The microdelta fronts are lobed and prograding. They occur as solitary bodies and do not occur in trains. They are 20–25 cm wide, 15–20 cm long and 5–10 cm thick.

Because of high water content in the mud, it was not possible to obtain a cross-sectional view of the microdeltas. It is, however, expected from their mode of formation that they too produce microdelta cross-laminations as a result of migration of the small delta with well-developed lee faces (McKee 1957; Jopling 1967). The cohesive nature of mud in these microdeltas imparts their different appearance from that of the sandy microdeltas. However, they exhibit transverse growth marks on their surface maintaining parallelism with their outermost lobes. Mud transported along the gently-sloping creek margins is responsible for microdelta formation. The pathway of mud transportation is initiated by gullies or rills.

5.3.3 Mud Pellets and Mud Lumps

Mudflats around creek margins are generally strewn with mud pellets and mud lumps (Fig. 5.2). The pellets are flat sub-elliptical bodies with axial lengths ranging 2–5 cm. Shape analysis of the pellets reveals that the pellets mostly plot on the prolate and oblate sectors in the Zingg's shape class (Fig. 5.3). The exposed portions of mud banks undergo erosion due to undercutting by tidal currents. As a result, chunks of mud of variable dimensions collapse from the mud banks. These mud chunks later break down into smaller pieces. These pieces of mud gradually

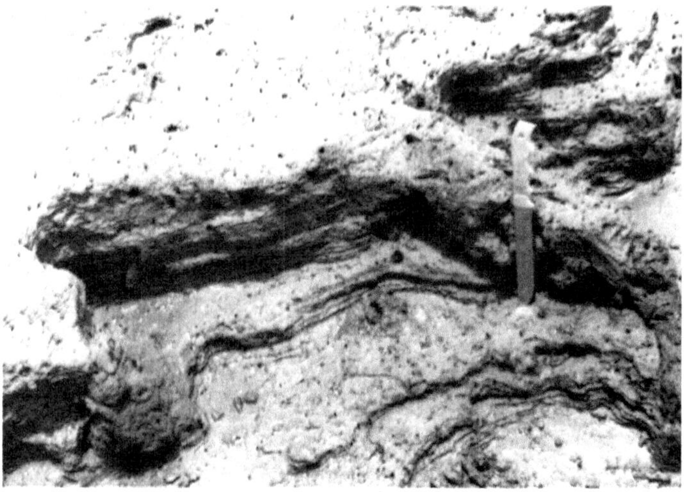

Fig. 5.2 Tidal creek margin showing exposure of longitudinal bedding (ECS) on the creek margin. Mud pellets are strewn on the muddy surface

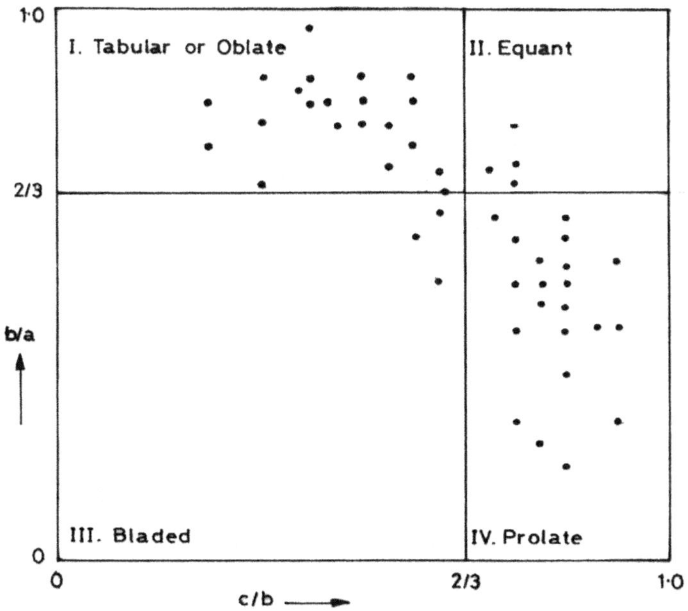

Fig. 5.3 Plots of mud pellets of Thakuran River bed on Zingg's shape class

become rounded by the to-and-fro movements of tidal currents and attain the pellet forms. They are not transported over very long distances because of the cohesive nature of mud. These pellets are distinctly different from those produced by biological activities of crustaceans and polychaetes in many modern tidal flats (Shinn et al. 1969).

5.3.4 Wood Clumps on Mudflats

Clumps of mangrove wood are often associated with muddy sediments. These wood trunks or stems, 1–3 mm long and 2–30 mm in diameter drift with tidal currents and get stuck into the mud at lower flow velocity. Their presence in the mudflat is a definite indication of their origin in a tropical environment. These wood trunks are derived from the eroded bank material sustaining mangrove bushes. Van Straaten (1961) described a similar association of wood clumps in the Dutch intertidal mudflats.

5.4 Stratification in Tidal Creek Deposits

The nature of stratification in the tidal creek deposits is generally difficult to identify from the sediments. But dried-up core samples brought into the laboratory revealed almost every detail of the lamination patterns. Generally the laminations are made up of alternating mud and fine sand. On many occasions, mm-scale thin laminae of fine sand form elongated lenses within the mud laminae. The types of laminae described in the following sections are recognisable.

5.4.1 Epsilon Cross-Stratification (ECS)

These are low-angle (5°–12°) cross-stratifications that extend as a single set over the whole thickness of the point bars bordering tidal creeks. They appear as inclined laminae (Allen 1963) on the margins of point bars, the upper surfaces of which dip equally to that of the laminae. The cross-strata dip normal to the current direction and range in thickness 1–3 m. The cross-stratifications are traceable laterally for over tens of metres.

Rahmani (1988) and Smith (1988) described epsilon cross-stratifications from the point bars of tidal rivers under mesotidal influence. The stratifications in ECS are rhythmic sand-mud couplets (Fig. 5.2). Sand is predominantly fine to very-fine grained, whereas, mud is predominantly silty. The laminae are of mm-scale in thickness. Thin lenses of wood fragments and organic litters are associated with the laminae and are derived from the mangrove vegetation. The laminae are profusely distributed by crab burrows and mangrove roots. Intensity of bioturbation increases upward where the mudflat appears with surficial pits and blisters. The rhythmicity in sand-mud laminae related to seasonal flooding of the tidal flats has also been described by Mossop et al. (1982) and Smith (1984).

5.4.2 Alternate Sand-Mud Sequence (Heterolithic Sequence)

A detailed study has been made of the lithologs from the core samples collected from the mudflat of the mid-channel bar of Paschim Sripatinagar. The sequence generally consists of mm-thick parallel laminae of sand and clay in regular and non-regular alternations. Occasionally the lamination thickness ranges up to a few cm. Lamina thickness diminishes upward; sand laminae reduces from 1 cm to 1 mm and mud laminae from 3 mm to 1 mm. This suggests a gradual decrease in current velocity in the upward direction. This alternation of sand–mud laminae resembles those described from recent tidal environments by Terwindt (1971) and Reineck and Singh (1980). Roep and Van Regteren Altena (1988) described similar features from palaeotidal environments. The lamination in the uppermost part of the sequence is crinkly due to interference by plant roots (Van Straaten 1954; Roep and Van Regteren Altena 1988).

A few core sections from mudflats adjoining tidal creeks expose bundle sequences with symmetrical and asymmetrical cycles of mm-scale, sand-clay laminae. The symmetrical cycles with 25–28 sand laminae are characterised by a gradual increase in thickness of sand laminae followed by a gradual decrease. This indicates their origin from neap-spring-neap cycles (Visser 1980; Reineck and Singh 1980). The asymmetrical cycles, on the other hand, exhibit increasing thickness of sand laminae and indicate deposition in neap-to-spring intervals and not in the spring-neap periods.

5.4.3 Convolute Lamination

Laminae convoluted or contorted into various degrees are noticed in the mud banks of tidal creeks. The laminae show gentle undulations (Fig. 5.4) to complicated folding or crumpling within non-deformed laminae. Of the several explanations proposed for the genesis of convolute laminations from different settings (McKee et al. 1962; McKee and Goldberg 1969; Reineck and Singh 1980), truncation of some of the complicated folded laminae by the overlying set of horizontal laminae indicates that the deformations took place in the unconsolidated state and the penecontemporaneous currents scoured the upper part of the folded layers.

Fig. 5.4 Convolute bedding in the muddy bank of a tidal creek. The deformations are restricted to a small unit indicating penecontemporaneous origin

5.5 Summary

Depositional features typical of tidal environment like mud couplets, tidal bedding and tidal bundles have been recognised in the tidal creek and mudflats of the Thakuran River. Both ebb-flood and neap-spring cycles have been established from the preserved physical sedimentary structures. The assemblage of sets of laminae corresponds to the bundle sequence of Visser (1980). Further the sequences are arranged in cyclic patterns and reflect spring-and-neap tidal origin. The semi-diurnal tides in this region with slight tidal asymmetry give rise to cyclicity in sedimentation (Visser 1980; Allen and Homewood 1984). The number of bundles in the cycles may not be precisely correlated with tidal range periodicity (28.5 bundles/cycle). Deviation in the number of bundles takes place due to the tropical cyclonic storms in the area and variations in the Ganges-Brahmaputra River discharge.

References

Allen JRL (1963) The classification of cross-stratified units with notes on their origin. Sedimentology 2:93–114

Allen PA, Homewood P (1984) Evolution and mechanics of Miocene tidal sandwave. Sedimentology 31:63–81

Das GK (2015) Estuarine morphodynamics of the Sunderbans. Coastal research library, vol 11. Springer, Switzerland, 211p

Faas RW (1981) Rheological characteristics of Rappahanock estuary muds, Southeastern Virginia, USA. Spec Publ Int Asso Sediment 5:505–515

Jopling AV (1967) Origin of laminae deposited by the movement of ripples along a stream bed. A laboratory study. J Geol 75:287–305

McKee ED (1957) Flume experiments on the production of stratification and cross stratification. J Sed Petrol 27:129–134

McKee ED, Reynolds MA, Baker CH Jr (1962) Laboratory studies on deformation on unconsolidated sediments. U S Geol Survey. Profess Papers 450-D:151–155

McKee ED, Goldberg M (1969) Experiments on formation of contorted structures in mud. Geol Soc Amer Bull 80:231–244

Mossop GD, Flach PD, Pemberfon GS, Hapuins JC (1982) Athabasca oil sands, sedimentary and development technology. Intern Assoc Sed Congr. MeMaster University, Hamilton, Ontario, Field Excursion 22, 59p

Rahmani RA (1988) Estuarine tidal channel and nearshore sedimentation of a late Cretaceous epicontinental sea, Drumheller, Alberta, Canada. In: deBoer PL, VanGelder A, Nio SD (eds) Tide-influenced sedimentary environments and facies. D Reidel Publishing Company, pp 433–471 (530p)

Reineck HE, Singh IB (1980) Depositional sedimentary environments. Springer, New York 408p

Roep TB, Van Regteren Altena JF (1988) Paleotidal levels in tidal sediments (3800 3635 BP). Compaction, sea level rise and human occupation (3275–2620 BP) at Bovenkarspel, N W Netherlands. In: deBoer PL, VanGelder A, Nio SD (eds) Tide-influenced sedimentary environments and facies. D Reidel Publishing Company, 530p

Shinn EA, Lloyd RM, Ginsburg RN (1969) Anatomy of a modern carbonate tidal flat, Andros island, Bahamas. J Sed Petrol 39:1202–1228

Smith DG (1988) Modern point bar deposits analogous to the Athabasca oil sands, Alberta, Canada. In: deBoer PL, VanGelder A, Nio SD (eds) Tide-influenced sedimentary environments and facies. D Reidel Publishing Company, pp 417–432 (530p)

Smith DG (1984) Vibracoring fluvial and deltaic sediments: tips on improving penetration and recovery. J Sed Petrol 54:660–683

Terwindt JHJ (1971) Litho-facies of inshore estuarine and tidal-inlet deposits. Geol en Mijnbouw 50:515–526

Van Straaten LMJU (1954) Sedimentology of recent tidal flat deposits and the parameters du Condroz (Devonian). Geol Mijnb 16:25–47

Van Straaten LMJU (1961) Sedimentation in tidal flats areas. J Alberta Soc Petrol Geol 9:203–213

Van Weering Tj CE (1981) Recent sediments and sediment transport in the northern North Sea: surface sediments of the Skagerrak. Spl Publ Int Asso Sediment 5:335–359

Visser MJ (1980) Neap-spring cycles reflected in Holocene subtidal large scale bedform deposits: a preliminary note. Geology 8:543–546

Chapter 6
Sediment Texture

Abstract Granulometric studies of sediment samples from different geomorphic units (e.g. mid-channel bars or flood-tidal delta, river mouth bars or ebb-tidal deltas, point bars, swash platforms, wash-over flats and riverbanks (levee) of the Thakuran basin have revealed the prevailing hydrodynamic condition at the time of their deposition. A coarsening of graphic mean size (M_Z), betterment of sorting (σ_1) and negative to positive skewness (SK_1) have been noticed, progressively from the seaward point and in the landward direction. Variations of these textural parameters are related to decreasing energy levels in the same direction. A bipartite-granulometric model of sedimentation, which has a sand-to-mud transition from the seaward point of the tidal creek and in the landward direction, has been discerned. This is in contrast to a tripartite-granulometric model typical of estuaries.

Keywords Texture · Grain size · Cumulative curves · Sorting · Skewness · Kurtosis · Mean size · Thakuran river · Sunderbans

The sediments of the Thakuran River are siliciclastic with sand-silt-clay as the chief constituent. Texturally, a bipartite model of sedimentation is present, which is marked by sand-to-mud facies from seaward point and in the landward direction. Cleaning of mud by high wave action, reworking of material by greater tidal amplitude and influx of sand by long shore currents have increased the proportion of sand in the seaward stretch of the riverbed (Das 2015). The flood current is able to push the sand inland up to a distance of 35–40 km, beyond which, further landward, mud constitutes the major portion of deposits. As the creeks are not fed by any freshwater from upland, the typical situation of a tripartite-grain-size model of estuaries does not hold well for the Thakuran River.

In all, 121 sediment samples were collected from different geomorphic zones, as well as from different sedimentary structures of both physical and biogenic origin. As per requirement, some samples were collected from the upper few centimetres. These samples actually represent the physical conditions of depositions for a short period of time prior to sampling, and these were compared with sediments after the beds were reworked by benthic animals.

G.K. Das, *Tidal Sedimentation of the Sunderban's Thakuran Basin*,
DOI 10.1007/978-3-319-44191-7_6

Interpretation of the grain-size frequency curve is based upon the pattern of curves and splitting of each curve into segments separated by the marked breaks or inflections (Folk and Ward 1957). It is generally supposed that each segment relates to a given mode of transport (Visher 1969; Tanner 1959). In fact, the cumulative curves are assumed to be composed of two or more overlapping Gaussian (normal) distributions (Spencer 1963; Tanner 1959) or of different truncated normal distributions (Spencer 1963; Visher 1969). Sagoe and Visher (1977) and Middleton (1976) have examined the relationship between the grain-size distribution and hydraulics in which the inflection points of the cumulative curves represent a specific change of hydraulic conditions.

6.1 Shape of the Cumulative Curves

The sediment samples; 121 in total, were collected over the entire length of the river from different geomorphological areas like point bars, mid-channel bars, swash bars, riverbanks and areas of other morpho-ecological interest, for the purpose of grain-size analysis. It was possible to group the grain-size–frequency curves into three categories depending on the shapes of cumulative curves, grain-size range and number of inflections present in each curve. Interpretations of the curves were done by applying Visher's method (Visher 1969) of grain-size analyses. Three different types of curves, obtained by synthesizing the size-frequency data, are depicted in Fig. 6.1. The representative curves reveal the following characteristics.

6.1.1 Type I: *Cumulative Curves Having Size Range Within 0.0–4.0 phi, i.e. Within Sand Sizes*

This type of cumulative curve represents samples collected from mid–channel bars, swash bars and point-bar surfaces. The curve is almost straight, i.e. lognormal when plotted on a log-probability graph paper with a modal value around 2.5 phi. All the curves constructed from these samples show breaks at: (i) 2 phi towards the coarser end having materials less than 10 %; and at (ii) 3 phi towards the finer end having materials less than 20 %. This type represents sediments mainly transported as intermittent suspension loads (Middleton 1976). The modal value is 2.5 phi. The sediments are well-sorted and the distributions are mostly non-skewed (Table 6.1). Individual curve of this type show excellent superimposition one over the other, indicating similar patterns (Figs. 6.2, 6.3 and 6.4).

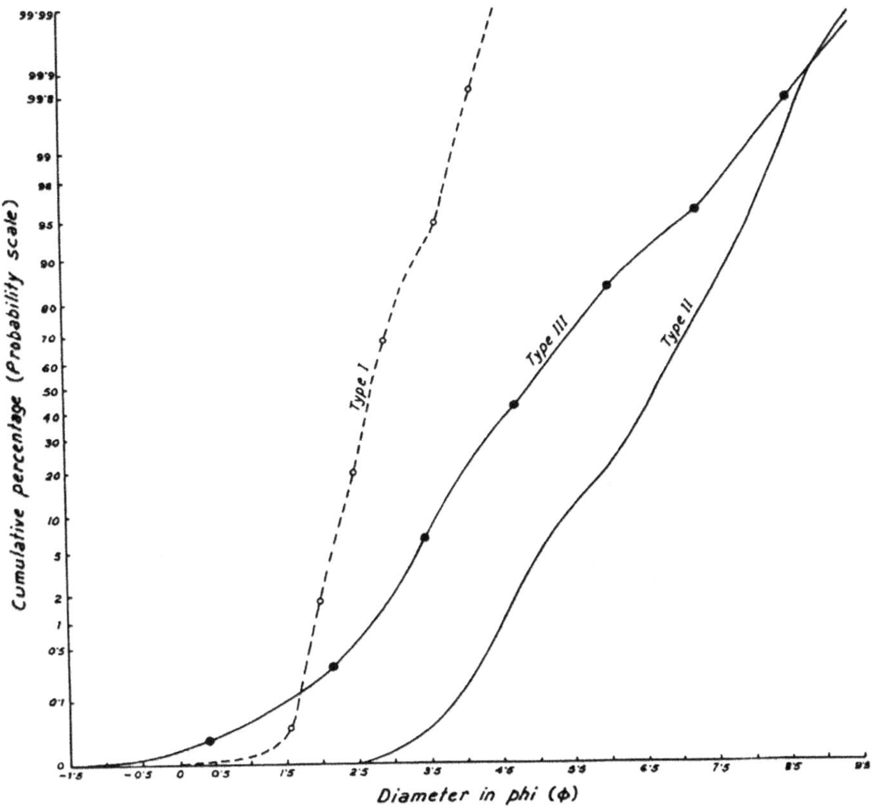

Fig. 6.1 Three representative grain size distribution curves of the Thakuran River samples

6.1.2 Type II: *Cumulative Curves Having Size Range Within 2.5–9.5 phi, i.e. Ranging from Fine-Sand to Clay Sizes*

The cumulative curve of this type represents samples collected from the upper portions of riverbank and mid-channel bars with muddy surfaces. The curve is slightly sinuous in pattern and deviates a bit from log normality when plotted on log-probability graph paper (Fig. 6.1). Grain size ranges from fine sand to clay (2.5–9.5 phi). A few distribution curves involve a wider range, from 2.5 to 11 phi (Fig. 6.5). A significant tailing off occurs towards the relatively coarser size, imparting a generally negatively-skewed character. Inflections occur at 5 and 7 phi. Materials coarser than 5 phi constitute less than 20 % whereas finer than 7 phi are less than 10 %. The modal value is at 5.5 phi. The sediments are mostly moderately

Table 6.1 Statistical size parameters and sand-silt-clay percentage of the sediment samples of the Thakuran River of the Sunderbans

1	2	3	4	5	6	7	8	9
Sample Locations (Distance in km from landward to seaward direction)	Sample no.	Graphic mean size (Mz)	Inclusive graphic standard deviation (σ_1)	Inclusive graphic skewness (SK_1)	Graphic kurtosis (K_G)	Sand %	Silt %	Clay %
Bhubankhali (10 km)	T_{70}	7.23	1.15	−0.27	0.92	3.20	96.80	–
Chuprijhara (14–16 km)	T_{67}	6.40	0.52	−0.17	1.76	1.10	98.90	–
	T_{68}	7.70	1.86	0.12	0.66	0.89	63.70	35.41
	T_{69}	6.65	0.62	−0.29	1.37	0.43	99.57	–
Purba Jatar Deul (19–20 km)	T_{65}	6.05	0.72	−0.33	1.34	0.88	99.12	–
Bhubaneswari Dwip (25 km)	B_1	6.0	0.99	−0.05	1.47	5.04	91.02	3.94
	B_2	6.43	0.66	−0.07	2.19	4.42	94.94	0.64
	B_3	6.31	0.71	−0.09	1.39	4.33	90.76	4.91
	B_4	6.39	0.69	−0.13	1.98	5.34	91.39	3.27
Damkal Dwip (30–32 km)	T_{32}	6.05	0.67	−0.29	1.06	4.38	93.59	2.03
	B_5	6.10	0.65	−0.21	0.97	5.37	92.01	2.62
	B_6	6.07	0.69	−0.23	1.02	5.81	91.39	2.80
	B_7	6.09	0.63	−0.31	0.99	6.01	92.47	1.52
Maipith (42 km)	T_{15}	5.97	0.94	−0.48	1.28	3.61	87.88	8.51
	T_{16}	5.92	0.87	−0.62	0.80	4.46	85.40	10.14
	T_1	5.95	0.73	−0.39	0.96	11.74	87.88	0.38
	T_2	5.90	0.84	−0.30	0.66	22.61	72.88	4.51
	T_3	5.96	0.96	−0.15	0.93	13.14	85.42	1.44
	T_4	5.77	0.63	−0.43	0.74	12.94	86.92	0.14
	T_5	6.30	1.08	−0.30	0.69	18.49	77.73	3.78

(continued)

Table 6.1 (continued)

1	2	3	4	5	6	7	8	9
Sample Locations (Distance in km from landward to seaward direction)	Sample no.	Graphic mean size (Mz)	Inclusive graphic standard deviation (σ_1)	Inclusive graphic skewness (SK_1)	Graphic kurtosis (K_G)	Sand %	Silt %	Clay %
	T_6	4.86	1.45	0.22	0.93	42.62	57.12	0.26
	T_7	5.80	0.79	−0.34	0.63	28.64	68.60	2.76
	*T_{17} (90 cm)	4.85	0.74	0.86	0.66	39.43	51.32	9.24
Paschim Sripatinagar north (42–44 km)	T_{33}	6.45	0.47	0.01	1.77	0.69	97.29	2.02
	T_{34}	2.02	0.22	0.50	1.31	99.98	0.02	–
	T_{35}	2.17	0.30	0.76	0.87	99.91	0.09	–
	T_{36}	2.32	0.36	0.13	0.86	98.32	1.68	–
	T_{37}	2.22	0.33	0.31	0.84	99.56	0.14	–
	T_8	3.55	0.84	0.09	2.52	85.01	14.29	0.70
	T_9	2.85	0.43	0.03	1.74	97.98	2.02	–
	*T_{10} (90 cm)	2.68	0.40	0.09	1.64	97.96	2.04	–
	T_{11}	2.32	0.31	0.11	0.97	100.0	–	–
	T_{12}	6.01	0.90	−0.25	1.72	7.0	90.36	2.64
	T_{13}	2.27	0.32	0.17	0.74	99.81	0.19	–
	T_{14}	2.78	0.37	0.21	1.93	95.61	4.39	–
	T_{38}	2.60	0.52	−0.06	0.72	97.80	2.20	–
	*T_{39} (30 cm)	2.56	0.57	0.80	1.35	93.69	6.31	–
	*T_{40} (60 cm)	2.65	0.55	0.17	1.60	92.03	7.97	–
	*T_{41} (90 cm)	2.63	0.52	0.13	1.20	93.07	6.93	–
	T_{42}	2.45	0.35	−0.10	0.90	100.0	–	–

(continued)

Table 6.1 (continued)

1	2	3	4	5	6	7	8	9
Sample Locations (Distance in km from landward to seaward direction)	Sample no.	Graphic mean size (Mz)	Inclusive graphic standard deviation (σ_1)	Inclusive graphic skewness (SK_1)	Graphic kurtosis (K_G)	Sand %	Silt %	Clay %
	$^*T_{43\ (30\ cm)}$	2.68	0.49	0.15	2.25	95.05	4.95	–
	$^*T_{44\ (60\ cm)}$	2.60	0.36	−0.12	1.38	100.0	–	–
	$^*T_{45\ (90\ cm)}$	2.65	0.38	−0.11	1.22	97.36	2.64	–
	T_{46}	2.56	0.34	−0.07	1.04	100.0	–	–
	$^*T_{47\ (30\ cm)}$	2.78	0.47	0.18	1.99	95.64	4.36	–
	$^*T_{48\ (60\ cm)}$	2.96	0.51	0.31	1.34	85.76	14.24	–
	$^*T_{49\ (90\ cm)}$	2.88	0.50	0.27	1.94	94.43	1.57	–
	$^*T_{50\ (120\ cm)}$	2.88	0.53	0.32	2.10	91.22	8.78	–
Paschim Sripatinagar north (44–46 km)	K_4	6.05	0.84	0.10	1.37	0.80	98.80	0.40
	K_5	6.76	1.40	−0.37	2.08	6.89	90.32	2.79
	K_6	6.31	0.87	0.45	2.16	1.32	94.48	4.20
	K_7	6.50	0.59	−0.18	1.93	3.44	95.44	1.12
	K_8	5.41	0.78	0.35	3.17	0.16	96.12	3.72
	$^*K_{10\ (120\ cm)}$	5.46	1.19	−0.69	0.97	18.12	78.12	3.76
	$^*K_{14\ (120\ cm)}$	7.76	1.35	−0.38	0.78	2.36	97.53	0.11
	$^*K_{15\ (120\ cm)}$	7.01	0.61	−0.41	1.53	1.36	94.88	3.76
	K_{16}	5.96	0.89	−0.69	0.91	4.39	94.78	0.43
	K_{17}	5.97	0.99	−0.41	1.68	3.75	92.30	4.05
	K_{18}	5.89	0.76	−0.33	0.78	1.24	97.13	1.73
	K_{19}	5.86	1.40	0.46	0.91	0.94	94.47	4.79
	K_{20}	5.76	0.64	−0.48	0.84	6.43	93.01	0.56
	K_{21}	5.81	0.43	0.64	0.43	18.16	81.43	0.41
	K_{22}	5.99	0.81	−0.36	0.67	3.18	88.94	7.88

(continued)

Table 6.1 (continued)

1	2	3	4	5	6	7	8	9
Sample Locations (Distance in km from landward to seaward direction)	Sample no.	Graphic mean size (Mz)	Inclusive graphic standard deviation (σ_1)	Inclusive graphic skewness (SK_1)	Graphic kurtosis (K_G)	Sand %	Silt %	Clay %
Achintyanagar (48 km)	P_1	6.30	0.52	−0.26	2.51	4.17	87.96	7.87
	P_2	5.58	0.76	−0.72	0.49	20.16	79.48	0.36
	P_3	6.40	0.99	−0.81	1.58	7.43	92.04	0.54
	P_4	6.32	0.56	−0.56	2.34	0.85	94.36	4.79
Upendranagar (49 km)	L_1	6.31	1.34	−0.69	0.70	6.44	93.20	0.36
	L_2	5.58	1.69	−0.66	0.64	24.24	72.12	3.64
	L_3	4.45	1.80	0.32	0.60	49.84	49.40	0.76
	L_4	2.58	0.32	−0.34	1.50	96.61	3.39	–
	L_5	4.66	1.64	0.32	0.68	49.48	50.16	–
	L_6	2.61	0.21	−0.22	−1.22	100.0	–	–
	$^*L_{8\ (90\ cm)}$	4.53	0.46	0.73	2.60	77.56	21.00	1.44
Rakhalpur (50 km)	T_{61}	2.72	0.35	−0.08	2.13	98.30	1.70	–
	$^*T_{62\ (30\ cm)}$	2.85	0.39	0.25	1.81	97.77	2.23	–
	$^*T_{63\ (60\ cm)}$	2.98	0.44	0.39	0.96	97.02	2.98	–
	$^*T_{64\ (90\ cm)}$	2.97	0.47	0.32	1.55	96.92	3.08	–
Sridharnagar (52–54 km)	T_{55}	2.80	0.25	0.15	1.64	98.84	1.16	–
	$^*T_{56\ (30\ cm)}$	2.75	0.31	0.06	1.71	98.67	1.33	–
	$^*T_{57\ (60\ cm)}$	2.73	0.31	0.18	1.41	98.59	1.41	–
	$^*T_{58\ (90\ cm)}$	2.76	0.31	0.11	1.77	97.92	2.08	–
	T_{59}	3.05	0.46	0.31	1.15	95.99	4.01	–
	T_{60}	2.63	0.32	0.00	1.40	98.94	1.06	–

(continued)

Table 6.1 (continued)

1	2	3	4	5	6	7	8	9
Sample Locations (Distance in km from landward to seaward direction)	Sample no.	Graphic mean size (Mz)	Inclusive graphic standard deviation (σ_1)	Inclusive graphic skewness (SK_1)	Graphic kurtosis (K_G)	Sand %	Silt %	Clay %
Chulkati II (54 km)	T_{25}	6.35	0.46	−0.18	1.73	1.58	98.42	–
Dhanchi north (57 km)	T_{51}	2.62	0.29	−0.12	1.57	98.99	1.01	–
	$^{*}T_{52\ (30\ cm)}$	2.72	0.47	0.46	1.89	97.94	2.06	–
	$^{*}T_{53\ (60\ cm)}$	2.82	0.42	0.27	1.69	97.53	2.47	–
	$^{*}T_{54\ (90\ cm)}$	2.77	0.40	0.11	2.04	97.22	2.78	–
Dhanchi (58 km)	D_4	2.92	0.40	0.24	1.37	98.28	1.72	–
	$^{*}D_{5\ (30\ cm)}$	2.23	0.53	0.43	0.92	98.16	1.84	–
	$^{*}D_{6\ (60\ cm)}$	6.27	0.47	−0.29	1.99	0.91	99.09	–
Dhanchi (60 km)	T_{22}	2.80	0.36	0.17	1.98	98.52	1.48	–
	$^{*}T_{23\ (60\ cm)}$	2.77	0.34	0.18	2.05	97.37	2.63	–
	$^{*}T_{24\ (90\ cm)}$	2.85	0.41	0.15	1.93	98.27	1.73	–
	D_1	6.53	0.48	0.06	1.45	–	100.0	–
	D_2	6.55	0.60	−0.16	1.57	1.24	98.76	–
	D_3	6.73	0.81	−0.02	1.13	0.40	99.28	0.32
Dhanchi (65 km)	T_{18}	2.83	0.35	0.14	2.04	97.88	2.12	–
	$^{*}T_{19\ (30\ cm)}$	2.92	0.42	0.21	1.69	97.53	2.47	–
	$^{*}T_{20\ (60\ cm)}$	2.98	0.39	0.19	2.05	96.84	3.16	–
	$^{*}T_{21\ (90\ cm)}$	3.02	0.45	0.16	1.27	96.38	3.62	–
Chulkati III (69 km)	T_{26}	5.87	0.78	−0.39	0.63	18.26	81.74	–

(continued)

Table 6.1 (continued)

1	2	3	4	5	6	7	8	9
Sample Locations (Distance in km from landward to seaward direction)	Sample no.	Graphic mean size (Mz)	Inclusive graphic standard deviation (σ_1)	Inclusive graphic skewness (SK_1)	Graphic kurtosis (K_G)	Sand %	Silt %	Clay %
Dhanchi south (73 km)	D_7	7.80	1.44	0.17	0.99	–	80.02	19.98
	D_8	6.42	0.55	−0.24	1.93	–	100.0	–
	D_9	6.47	1.43	0.12	0.91	–	95.89	4.11
	D_{10}	6.28	0.61	−0.35	1.88	2.91	96.96	0.13
	D_{11}	2.97	0.54	0.23	1.17	96.84	3.16	–
	D_{12}	2.82	0.36	0.14	2.04	98.60	1.40	–
	D_{13}	2.80	0.40	0.17	1.87	98.67	1.33	–
	D_{14}	2.75	0.69	0.14	1.09	96.94	3.06	–
	D_{15}	2.67	0.43	−0.02	1.74	98.31	1.69	–
	*$D_{16\ (30\ cm)}$	2.81	0.41	0.05	1.69	98.55	1.45	–
	*$D_{17\ (60\ cm)}$	2.62	0.36	−0.06	1.65	98.49	1.51	–
	*$D_{18\ (90\ cm)}$	2.92	0.63	0.11	1.05	95.80	4.20	–
Bulchery Island (74 km)	T_{27}	2.53	0.49	−0.15	1.99	98.01	1.99	–
	T_{28}	2.67	0.21	−0.06	1.64	99.47	0.53	–
	T_{29}	2.77	0.18	0.15	0.98	100.0	–	–
Bulchery south (80 km)	T_{30}	5.85	0.75	−0.44	0.60	6.45	92.50	1.05
	T_{31}	5.82	0.81	−0.29	0.63	22.63	75.34	2.03

*Asterisks for depth samples: depth of collection within parenthesis

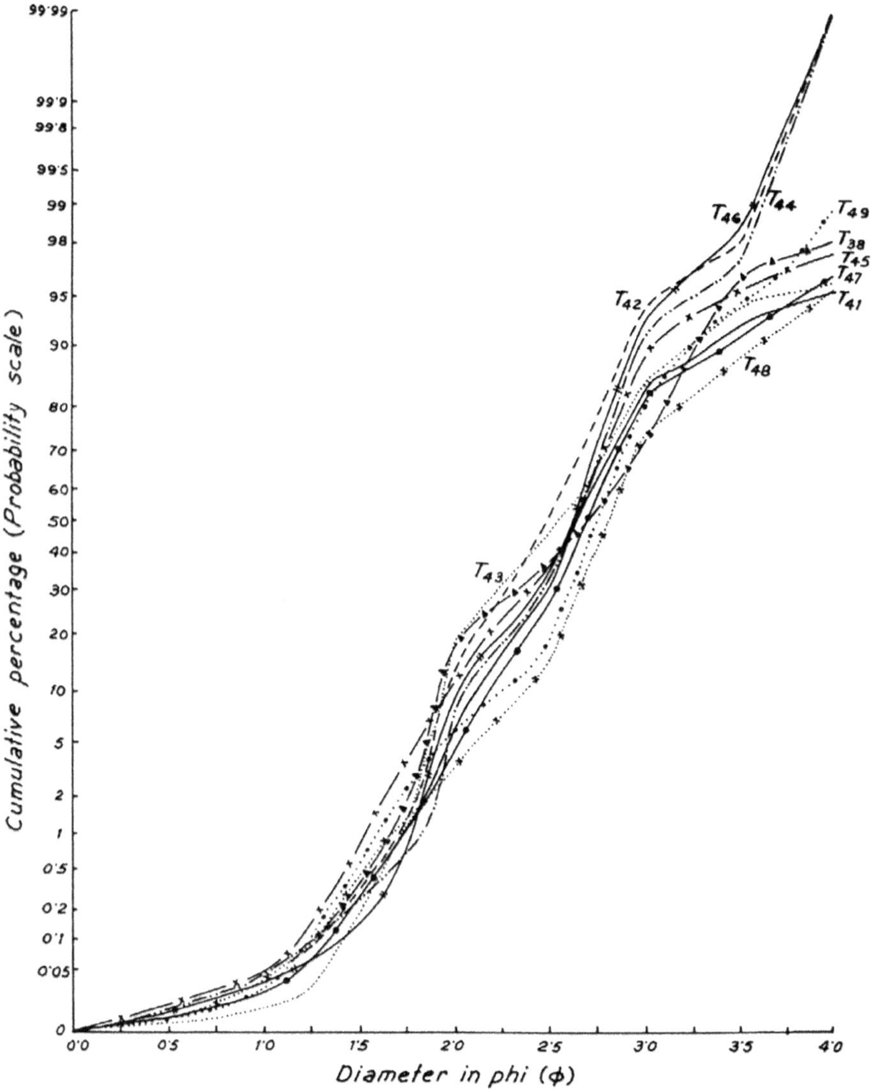

Fig. 6.2 Cumulative curves (Type I) of Paschim Sripatinagar mid-channel bar sand

sorted. The curves represent sediments mainly transported as intermittent suspension loads, with a more pronounced tailing off towards the coarser fraction (Table 6.2). Cumulative curves constructed from the samples show a unique similarity in pattern and superimpose one another when plotted on the log-probability graph paper (Figs. 6.5, 6.6, 6.7 and 6.8).

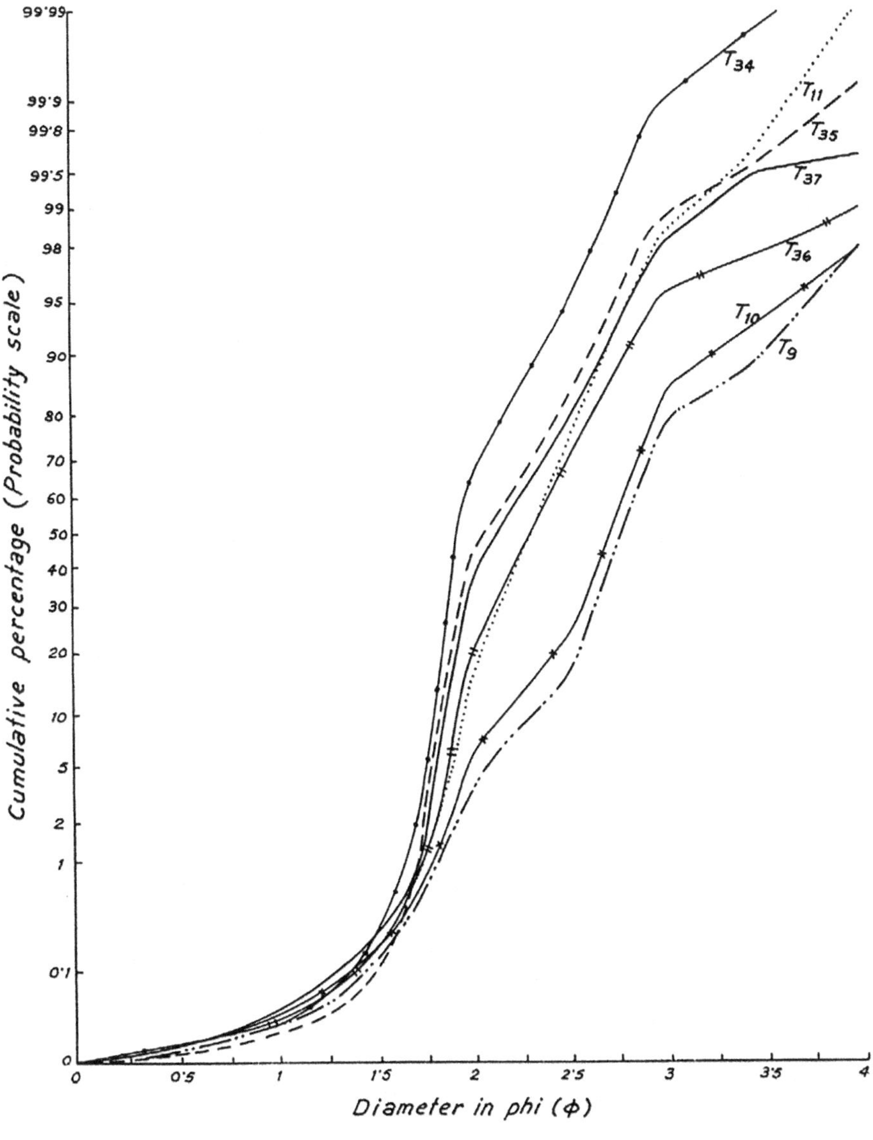

Fig. 6.3 Cumulative curves (Type I) of Paschim Sripatinagar mid-channel bar sand

6.1.3 Type III: *Cumulative Curves Having Size Range −1.0–9.0 phi, i.e. Ranging from Very Coarse-Sand to Clay Sizes*

The cumulative curve of type III represents samples collected from lower portions of riverbanks and point-bar margins. The curve forms a slightly zigzag pattern and

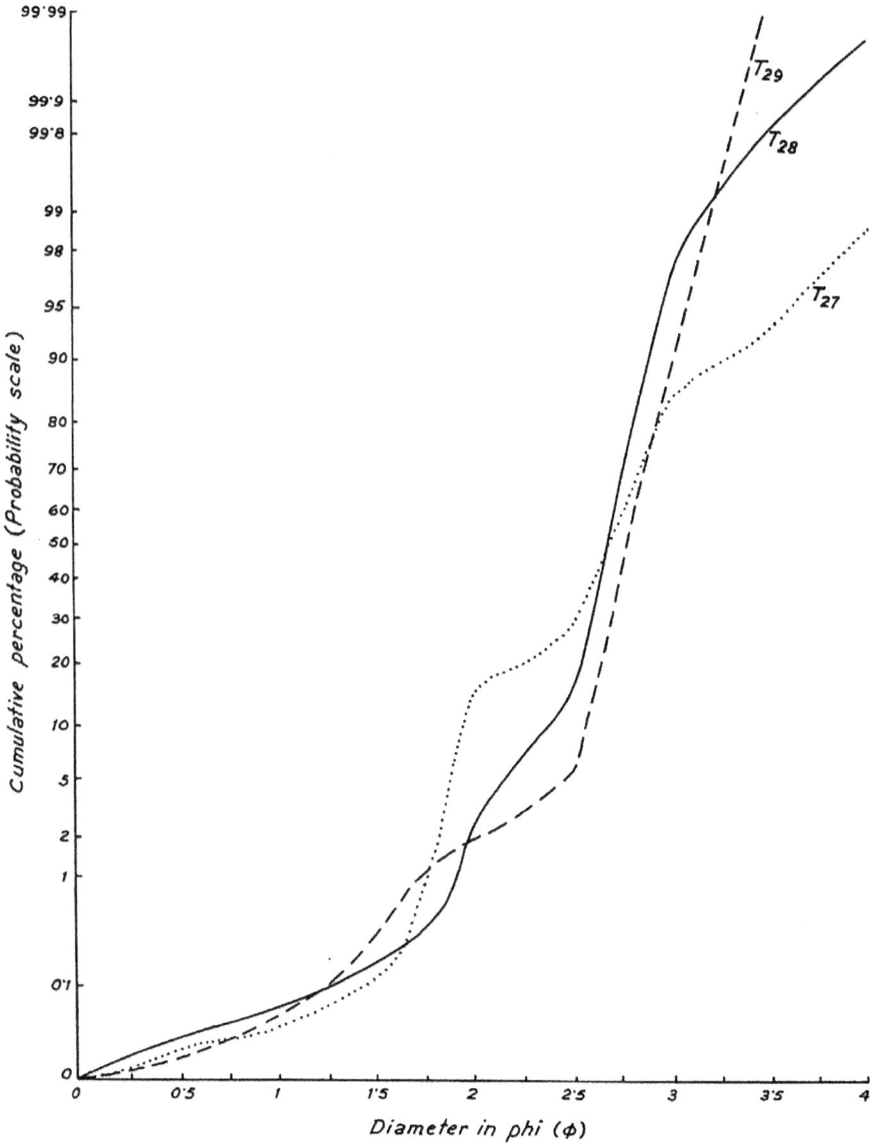

Fig. 6.4 Cumulative curves (Type I) of Bulchery point bar sand

reflects a non-log normal distribution when plotted on log-probability graph paper (Fig. 6.1). Grain size ranges from −1.0 to 9.5 phi. A more significant tailing off occurs towards the finer end of the curves giving, them a positively skewed nature. The inflections are noted at 2 and 6 phi. Materials coarser than 2 phi constitute less than 1 %, whereas, below 6 phi constitute greater than 50 %. The modal value is at

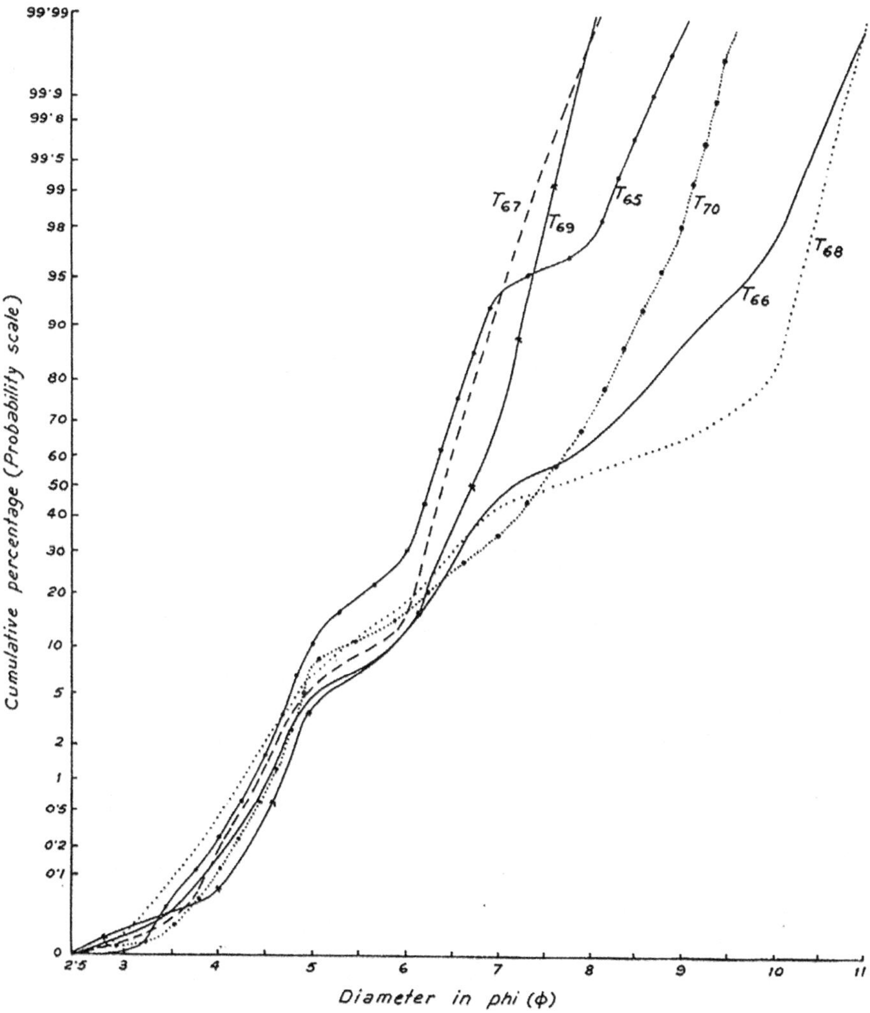

Fig. 6.5 Cumulative curves (Type II) of muddy bank samples between Bhubankhali and Jata

5 phi (Table 6.1). Sorting is moderate to poor but poorer than for type I and type II sediments. The tailing off towards the fine is a reflection of the admixture of finer suspended load to population of the intermittent-suspension sand load. All curves show a nice similarity of pattern and almost superimpose when plotted together (Figs. 6.9 and 6.10).

Table 6.2 Characteristics of the three categories of grain-size distribution curves showing inflections and sub-population patterns

	Type I	Type II	Type III
Grain size range	Sandy (0–4 phi)	Fine sandy to clayey (2.5–9.5 phi)	Sandy to clayey (−1.5–9.5 phi)
Inflection (phi)	2–3	5–7	2–6
Amount of coarser sub-population (%)	10	20	1
Amount of finer sub-population	20	10	50
Modal value (phi)	2.5	6.5	5
Sorting (σ_1)	Well sorted	Moderately sorted	Moderately to poorly sorted
Skewness (SK_1)	Non-skewed (−0.1 to +0.1)	Mostly slightly negatively skewed (−0.16 to 0.69)	Positively skewed (+0.22 to +0.45)
Principal mode of transport	Intermittent suspension load	Intermittent suspended load with minor admixture of suspended load	Intermittent suspended load with a huge admixture of suspended load

6.2 Inter-relationship Between Grain Size and Morphological Units

The different units have distinctiveness in terms of their grainsize characteristics, which, in turn, reflect variation in hydrodynamic conditions.

It is evident from the correlation table between grain-size distribution and geomorphologic areas of sampling (Table 6.3) that grains deposited from the intermittent suspension point to the prevalence of higher-energy conditions in the upper portions of the point bar and mid-channel bar (Type I). Grains deposited from intermittent suspension along with a bulk of finer suspended load reflect a traction-cum fall-out deposition in a very low-energy condition in the upper portions of the riverbank (Type II). Type III, which has the poorest sorting and suspension settling of both fine sand and mud, indicates an intermediate hydrodynamic condition in the lower flanks of point bars and mid-channel bars.

6.3 Classification of Sediment Types

Folk (1954) proposed a triangular diagram to classify sedimentary rocks, but its application for the classification of estuarine or tidal sediments was not tested. Shepard (1954) proposed a triangular diagram to classify different sediment types

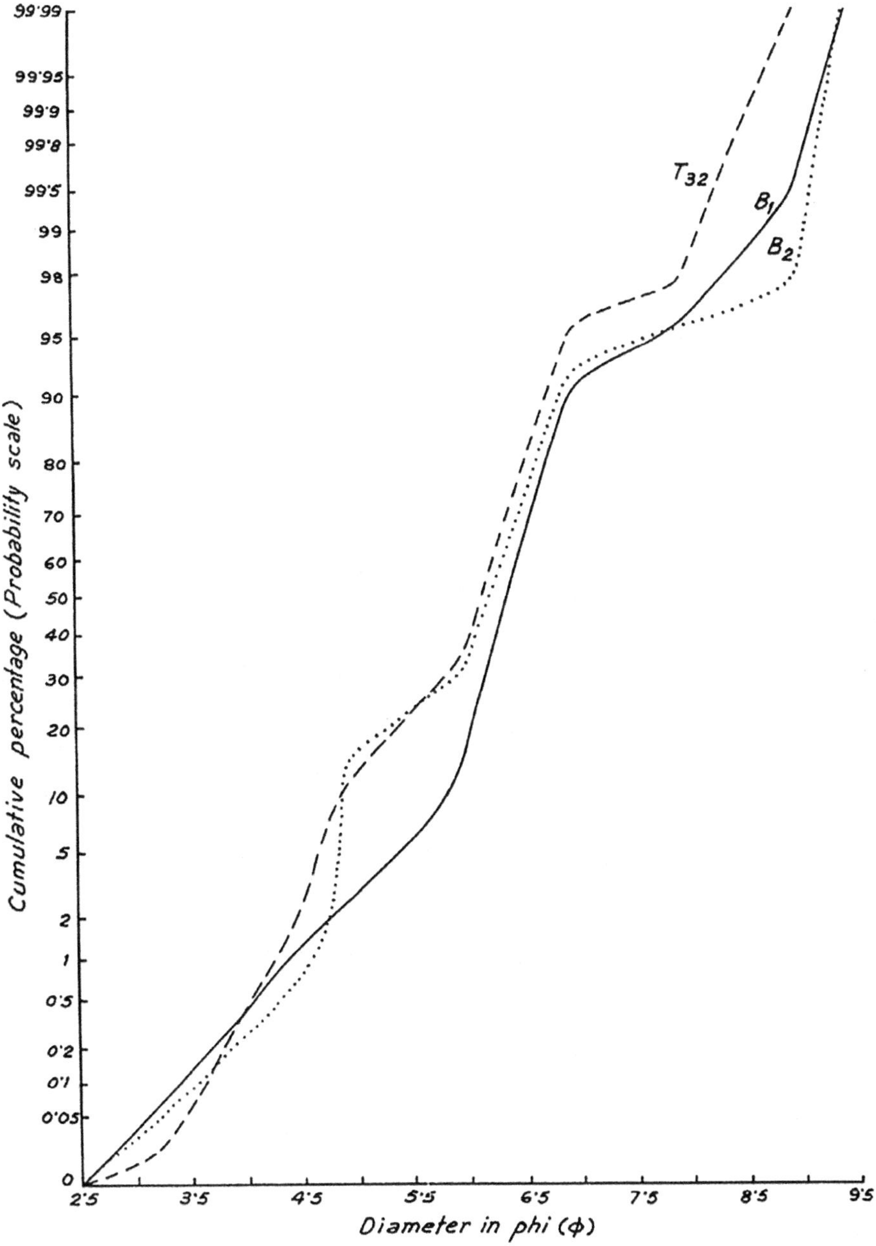

Fig. 6.6 Cumulative curves (Type II) of the mid-channel bar mud samples of Bhubaneswari (B1 and B2) and Damkal (T32)

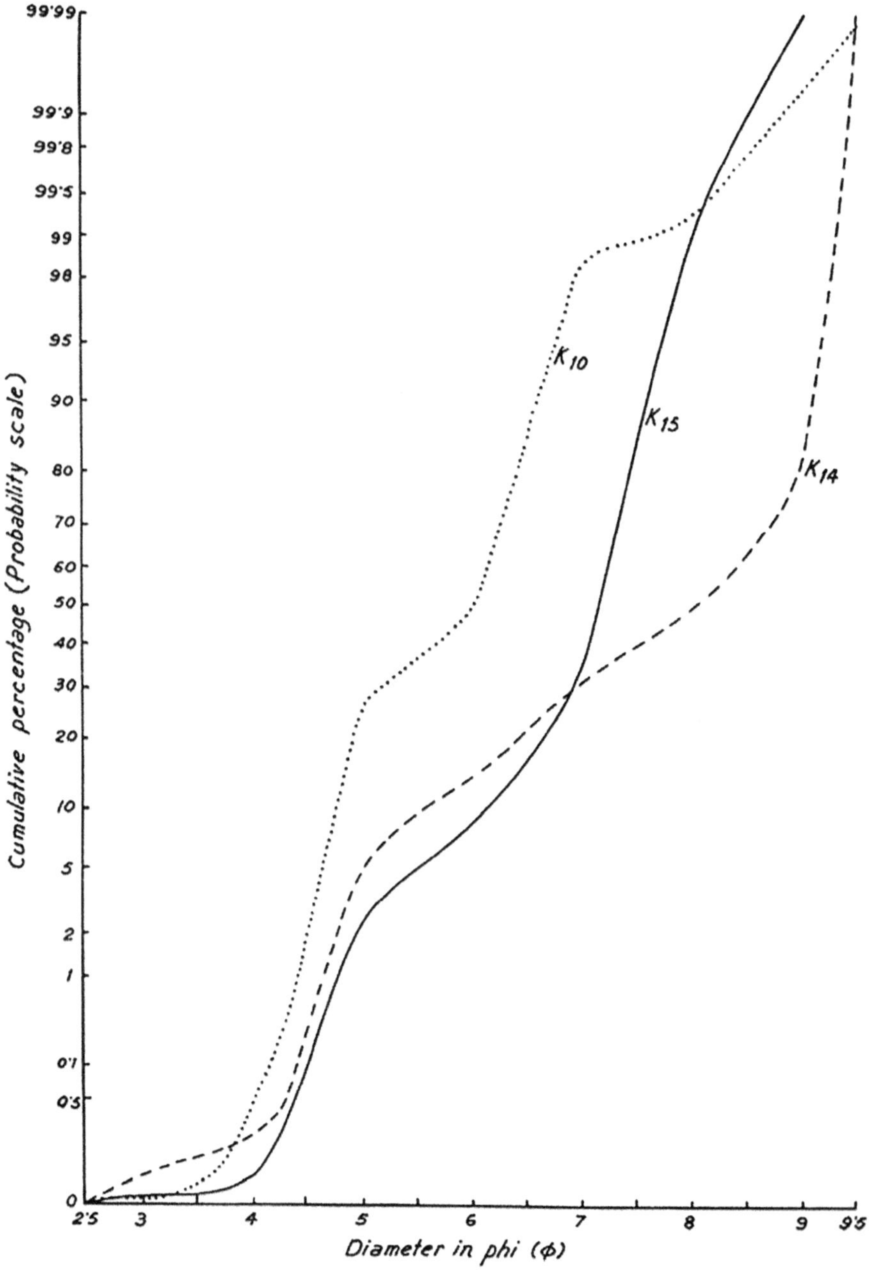

Fig. 6.7 Cumulative curves (Type II) of muddy bank samples around Paschim Sripatinagar

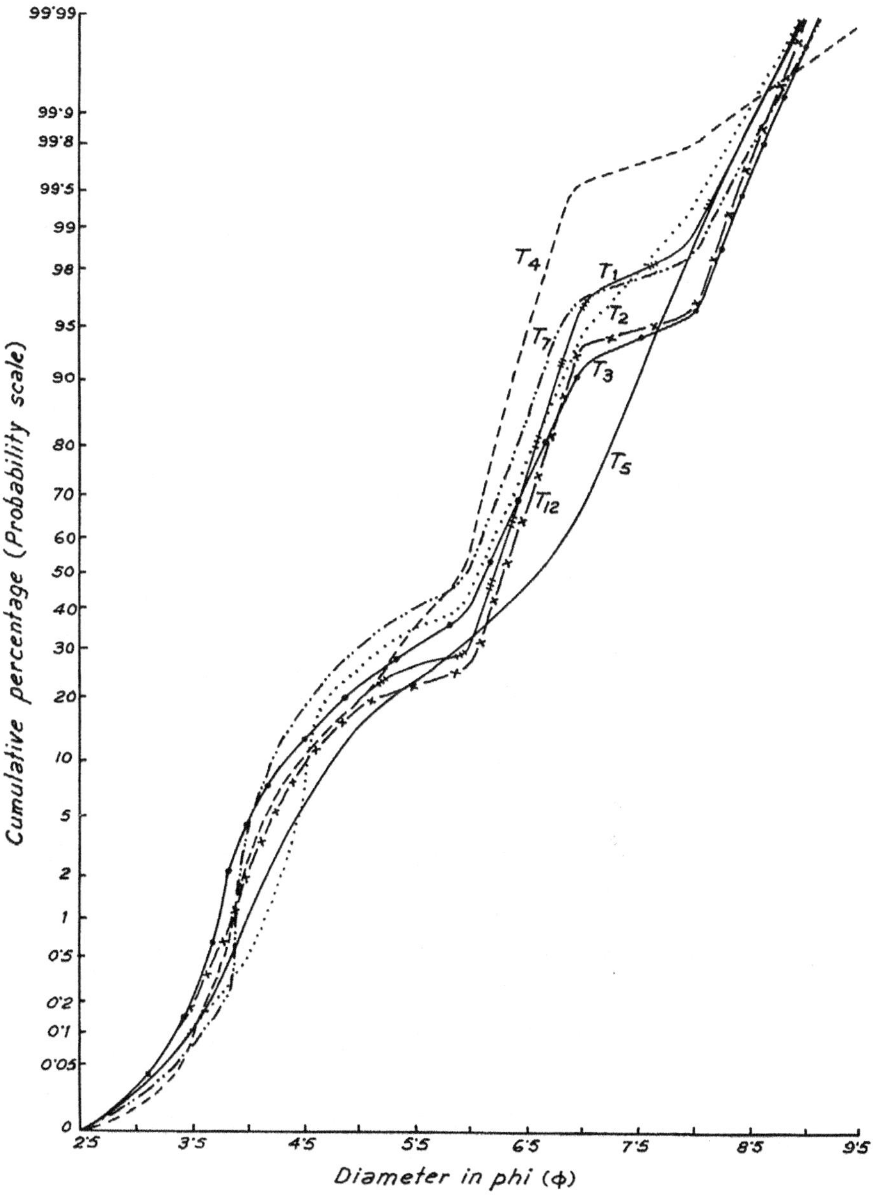

Fig. 6.8 Cumulative curves (Type II) of muddy samples of Paschim Sripatinagar mid channel bar

and to distinguish different sedimentary facies in estuarine environments (Pejrup 1988; Evans 1965; Flemming 1999). The sand-silt-clay percent of each sample of the present study area has also been plotted in Shepard's triangular diagram in order to classify the different sediment types.

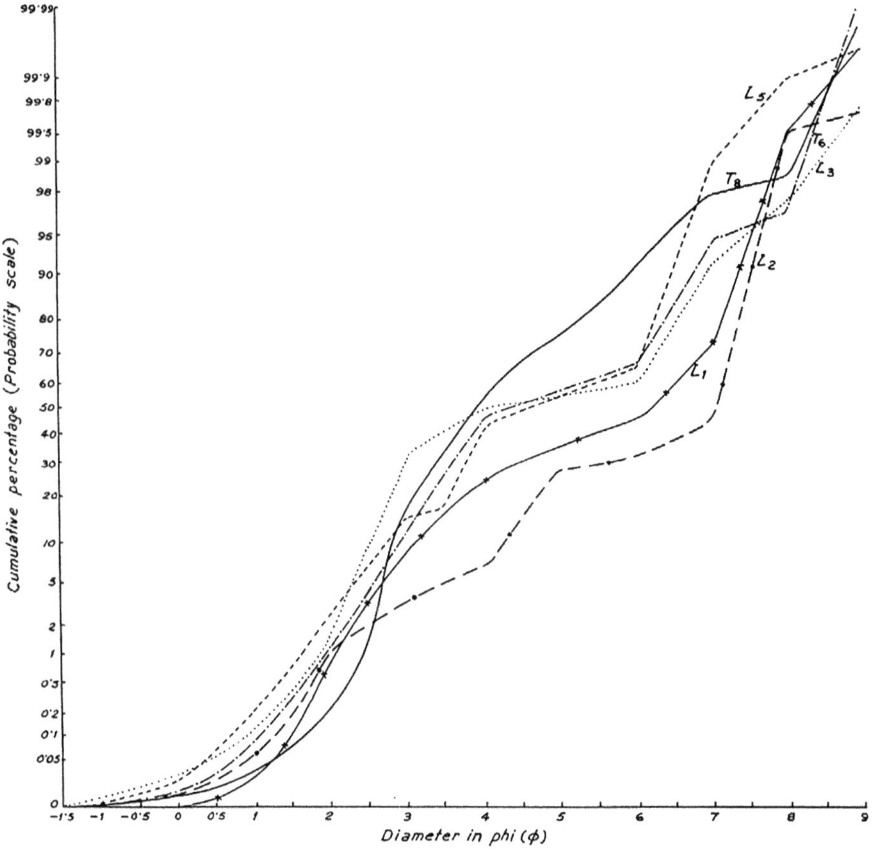

Fig. 6.9 Cumulative curves (Type III) of sandy to muddy point bar samples at and around Upendranagar

The cluster of plots primarily lies within a sand and silt field (Fig. 6.11a). The cluster in the sand field mostly denotes samples collected from the river close to the seaface, particularly from the mid-channel bar and swash platform. The concentration of plots in the silt field denotes samples collected from the river more towards the landward direction, particularly from the bank and also from point bars and mudflats adjoining the tidal creeks. The silty sand and sandy silt samples result from differential mixing of sand and silt fractions and signify the transitional middle stretch of the river.

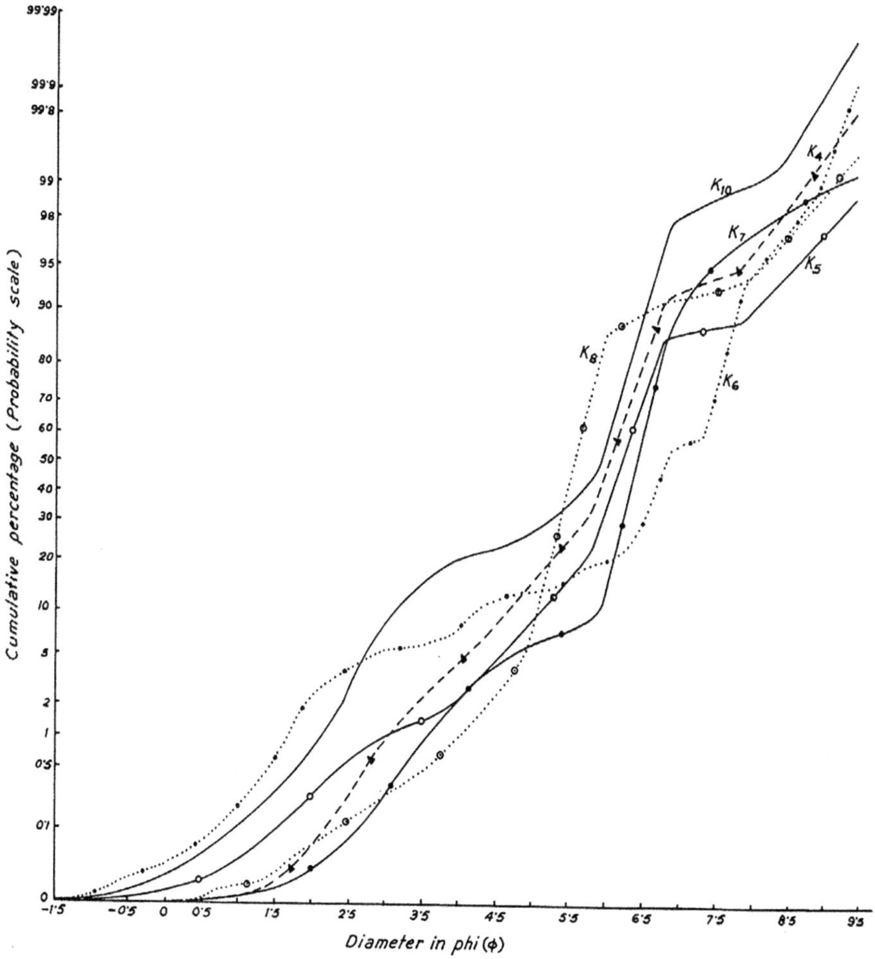

Fig. 6.10 Cumulative curves (Type III) of sandy to muddy river bank samples at and around Paschim Sripatinagar

Table 6.3 Correlation between types of grain-size distribution and geomorphological areas of sampling

Grain size distribution type	Upper portions of mid-channel bar, swash bar and point bar	Lower portions of mid-channel bar and point bar	Bank	
			Lower	Upper
Type I	X	–	–	–
Type II	–	–	–	X
Type III	–	X	X	–

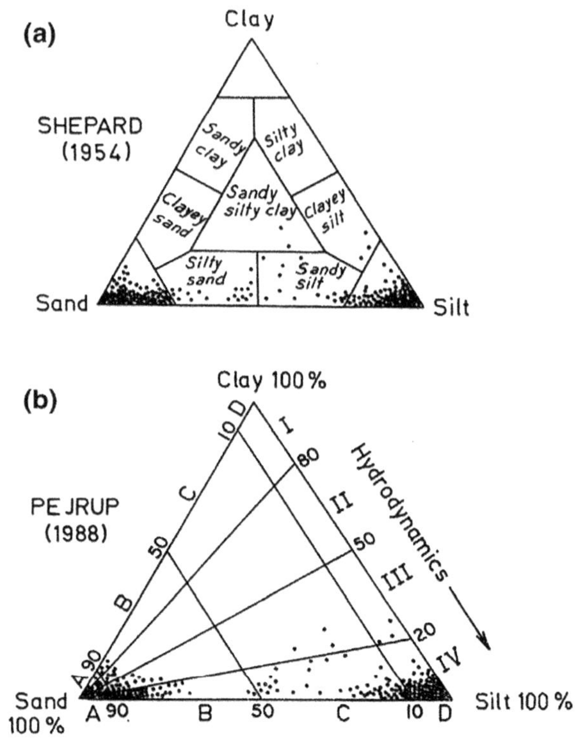

Fig. 6.11 **a** Plot of sand-silt-clay percentage of samples in the triangular diagram of Shepard (1954). **b** Plottings of the same samples in the hydrodynamic fences of Pejrup (1988)

6.4 Sediment-Trend Matrix of Grain-Size Relations

A sediment-trend matrix of grainsizes relation for the three types of grain-size distributions is given in Table 6.4. McLaren (1981) showed that such a matrix helps in identifying the possible source of a given deposit in a system-related environment. The sediment samples collected from the different geomorphic zones of the Thakuran River networks of the Sunderbans represent a gradation from one another as the materials belong to closely related sub-environments. McLaren (1981) suggested that, if a source-sediment is subjected to transportation, its original

Table 6.4 Composition of statistical size parameters for the three grain-size distribution types

Sediment source	Deposits		
	Size range	Inclusive graphic standard deviation (σ_1)	Inclusive graphic skewness (SK_1)
Type II	Sand-silt	Moderate	−ve
Type I	Sand	Well	Non-skewed
Type III	Sand-clay	Moderate to poor	+ve

grain-size distribution will be modified in one of the following ways (for a total deposition of sediment in transport):

1. The deposit is finer, moderate to well-sorted and more negatively skewed than the source sediment.
2. The lag sediment is coarser, better sorted and non-skewed to positively skewed than the source sediment.
3. The selective deposit from a sediment in transport can be either finer or coarser with better sorting and more positive skewness than that of the source sediments.

Applying the reasoning of McLaren (1981) to the Thakuran River sediments of the Sunderbans, it appears that sediment of Type II can be considered as a source for sediment of Type I and again the sediment of Type I can be possible source of Type III (Table 6.4). Removal or truncation of the silt fraction from Type II sediments changes the sediments to Type I so that sediments lie within sand sizes with better sorting and non-skewed nature. A stronger water current in the source area is responsible for producing such a change. Again, admixture of clay and sand in a relatively sheltered area brings the Type II sediments to Type III with moderate to poorer sorting and positive skewness.

6.5 Sediment—Hydrodynamic Relation

The hydrodynamic conditions for the depositional environment of the Thakuran drainage basin have been worked out by plotting the percentage of sand, silt and clay of all the samples in the triangular diagram of Pejrup (1988). According to Pejrup (1988), the hydrodynamic conditions for the depositional environment of an estuary or tidal creek may be described by the percentage of clay in the mud fraction. Sections I to IV within the triangle mark the increasingly violent hydrodynamic conditions (Fig. 6.11b). The sediments are classified further on the basis of their sand content into four sections from A to D. Thus the triangle has 16 divisions, each labelled by a letter and a number, such as A I–A IV; B I–B IV; C I–C IV and D I–D IV. The division A I represents sediments having 90 to 100 % sand deposited under high hydrodynamic conditions. All other divisions from A II to A IV having a specified range of sand content characterise the deposition under decreasing hydrodynamic conditions.

Mid-channel-bar, point-bar and swash-bar sandy samples generally fall into section IV (as A IV, B IV, C IV and D IV) and indicate strong hydrodynamic conditions of deposition. Again, sediment samples from the riverbank and contiguous swamp have a slightly higher content of clay (D III) and represent a quieter hydrodynamic condition of deposition. Some samples, however, plot in between sections III and IV. Variations in the sand content of sediments are caused by the changes in the differentiation trend from primary (Type II) to secondary (Type I) and then to tertiary (Type III), depending on the distances of transport and segregations involved therein (Table 6.3).

As a matter of fact, the grain size of the Thakuran River sediments nicely reflects the nature of source sediments and their hydrodynamic conditions of deposition. Generally, erosion prevails towards the seaward reach of the river with high wave energy, and deposition dominates in the landward reaches of relatively quieter environment. Thus, finer muddy sediments are deposited on the riverbanks, flanks of the mid-channel bars and point bars with low depositional energy. The coastal ridges off the Belgian coast also show similar segregation of finer from coarser fractions (DeMaeyer and Wartel 1988).

6.6 A Bipartite Model of Grain-Size Distribution

An overall bipartite (mud-to-sand) grain size distribution characterises the Thakuran River sediments of the Sunderbans (Fig. 6.12b) in contrast to the general tripartite (sand-to-mud-to-sand) distribution (Rahmani 1988) typical for estuaries (Fig. 6.12a). Rivers of the Sunderbans do not have any freshwater source upland. They behave like large tidal creeks. Their upland sediments are predominantly muddy, whereas the sediments near the funnel opening belong to the sand-grade. For an example, a clear transition from sand to mud occurs at a distance of 40–42 km upstream on the Thakuran River from its confluence with the Bay of Bengal. A sharp shift in the modal class of the sediments in this stretch of the river near Paschim Sripatinagar is clearly seen from the histograms representing grain-size distributions (Fig. 6.13). The median value of the muddy samples lies at 6.5 phi whereas it is 2.5 phi for the sandy samples. Elimination of mud seaward is due to higher wave action, greater tidal amplitude and the influx of clear sand from the mouth. Eventually on the basis of grain-size, the river can be divided into two major sedimentary provinces: a predominantly sandy province occupying the seaward 40 km and a finer silty to clayey province on the upper (northern) 40 km. The transition zone is marked by the intermingling of the two sediment provinces (Fig. 6.12).

The bipartite model of Thakuran without a true fluvial stretch of headwater supply is a distinct deviation from the condition of the common tripartite grain-size variation in many macrotidal-mesotidal estuaries (e.g. the Hugli estuary). In such estuaries, sand is brought in by flooding tides from nearby beaches and from near-shores exposures for a distance of at least a few km upstream. At this point, along with salt-wedge formation, the saline tidal water encounters the fresh fluvial, sediment-laden water and creates the very well-documented zone of maximum turbidity with thick mud deposits. Upstream of the muddy zone, river-derived sand is deposited. Lengths of these sandy and muddy reaches vary with the changes in the balance between tidal and fluvial processes.

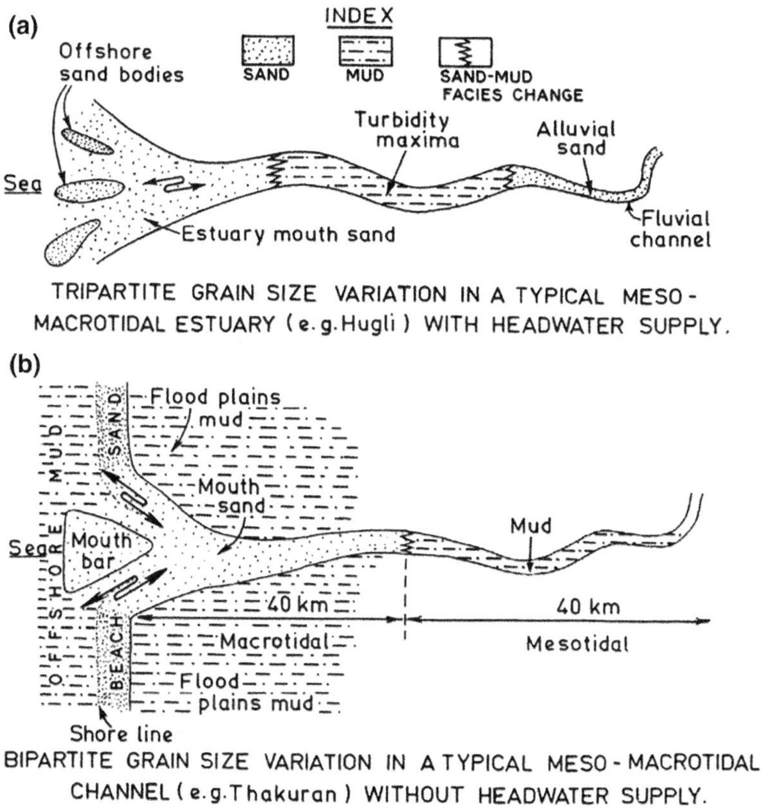

Fig. 6.12 An idealised diagram showing tripartite (**a**) and bipartite (**b**) grain size variations

The Thakuran River lacking any headwater supply does not possess the uppermost sand-filled portion. Here, in this river, the transition between the sand and mud facies occurs approximately 40 km upstream from the seaface. In this stretch of the river, the Paschim Sripatinagar mid-channel bar (4.5 km long at low tide) registers the mud (clayey silt) to sand transition. The northern half of the bar is generally muddy, whereas the southern half is predominantly sandy.

The lower reach of sand-filled estuaries (e.g. Hugli) or tidal creeks (e.g. Thakuran) terminating against muddy zones on both sides (tidal flats) can be easily mistaken for the shore-parallel-barrier sand that terminates in a muddy zone in both seaward and landward directions. A misinterpretation of this kind may result in a 90° error in mapping the local paleo-shoreline. The correct interpretation of this sand, having been deposited in a tidal estuary, perpendicular to the shoreline, may help to avoid such a serious error (Rahmani 1988).

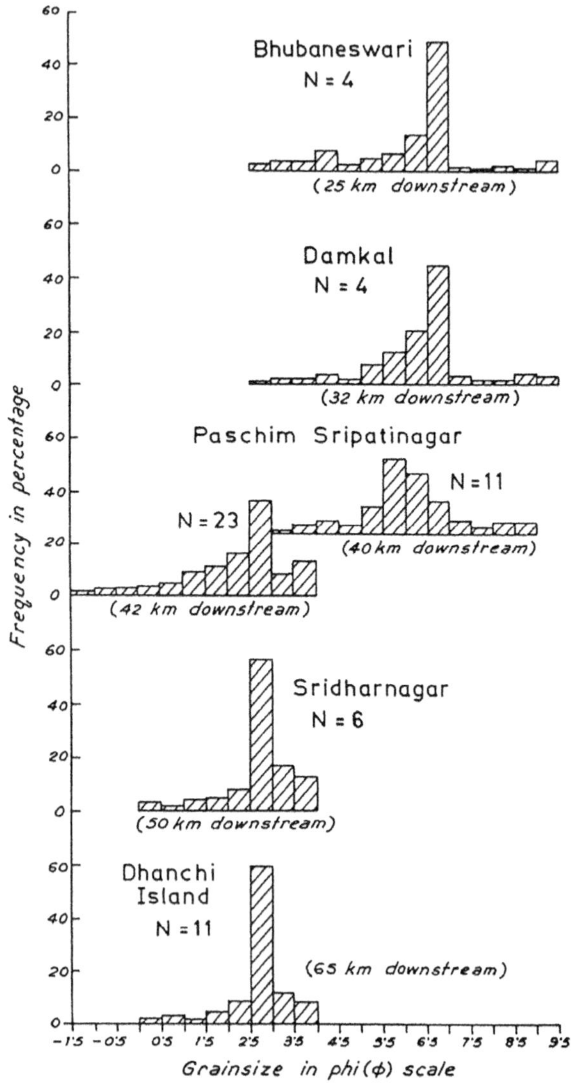

Fig. 6.13 Composite histogram of grain size distributions of the mid channel bar samples along river distance

6.7 Interrelationship of Grain-Size Parameters and River Distance

An attempt has been made to show the interrelationship of grain-size parameters and distance from the landward end of the river. Plots of graphic mean size (M_Z), inclusive graphic standard deviation (σ_1) and inclusive graphic skewness (SK_1) of the mid-channel bar and point-bar sediments against river distance (Figs. 6.14 and 6.15) display the following characteristics:

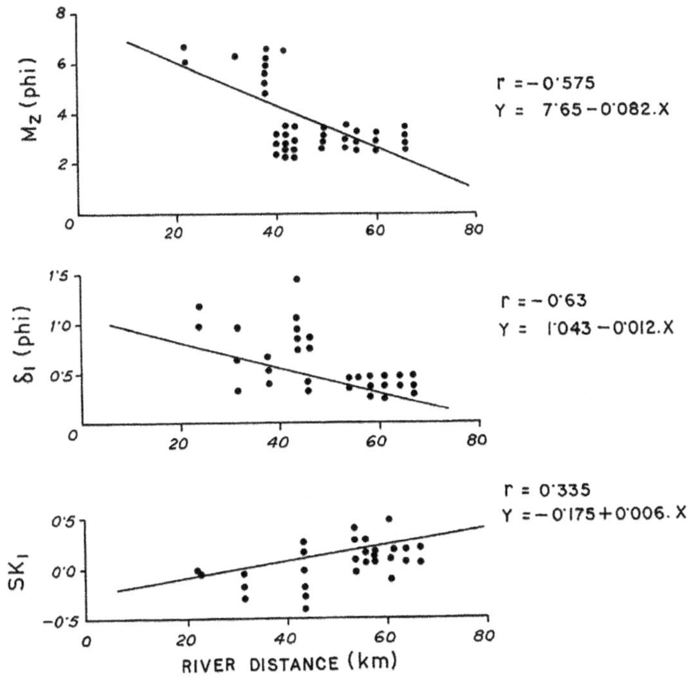

Fig. 6.14 Plot of Mean size (M_Z), Sorting (σ_1), Skewness (SK_1) of mid channel bar samples versus river distance

(i) Mean size (M_Z) of the samples of the mid-channel bar and point bar show a distinct seaward coarsening from silt to sand, inferring a negative linear correlation (r = −0.575 and −0.59 respectively).

(ii) Sorting (σ_1) of the sediments indicates betterment along the seaward direction. This is because of greater reworking of the sediments towards the seaface. The relation between sorting and river distance shows a negative correlation (r = −0.63 and −0.57 respectively).

(iii) The plot of skewness (SK_1) versus river distance shows slightly negatively skewed to non-skewed nature barring a few exceptions. This change in skewness reflects a change in grain-size distribution patterns from Type II to Type I (Table 6.2). This negative skewness of the landward sediments is caused by the minor admixture of clay to the intermittent suspended load. The sediments near the seaface are almost non-skewed and reflect a log-normal distribution without any significant admixture of coarser or finer subpopulations.

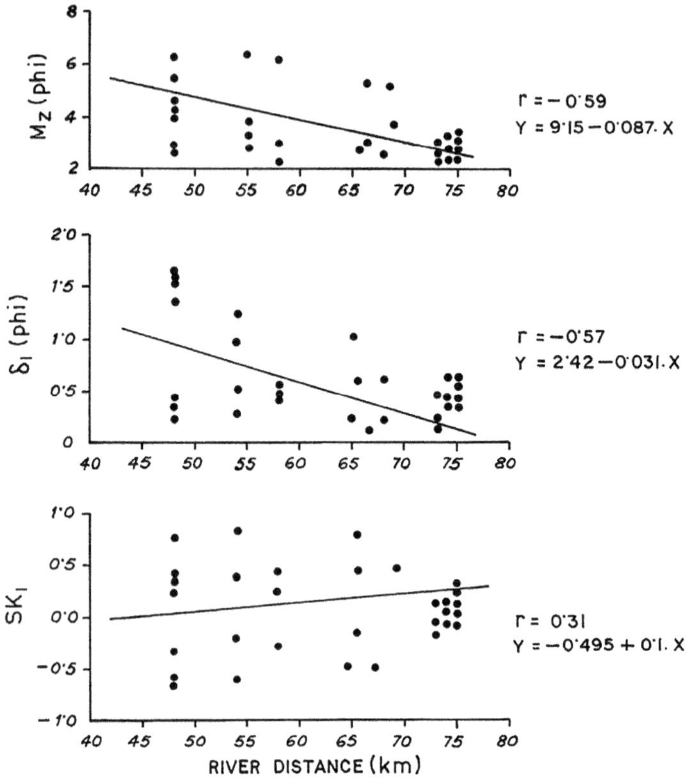

Fig. 6.15 Plot of Mean size (M_Z), Sorting (σ_1), Skewness (SK_1) of mid channel bar samples versus river distance

6.8 Environment Sensitiveness of Size Parameters

Environment sensitiveness of the statistical grain-size parameters was also tested following the classical work of Friedman (1967) and Moiola and Weiser (1968). The plots of graphic mean size (M_Z) versus inclusive graphic standard deviation (σ_1) for the total samples chiefly fall within the river environment and are distinctively different from the beach environment (Fig. 6.16a). Further, plots of the inclusive graphic standard deviation (σ_1) versus inclusive graphic skewness (SK_1) for the total samples also fall within the field for the river environment (Fig. 6.16b) as demarcated by Friedman (1967) and Moiola and Weiser (1968). The fields for river and beach as demarcated by Moiola and Weiser (1968) are a better fit for this plotting than that demarcated by Friedman (1967).

Testing of the environmental sensitiveness of grain-size parameters of the Thakuran River sediments of the Sunderbans helps in understanding a very significant fact. It is clear that, although the Thakuran River is a tide-dominated large

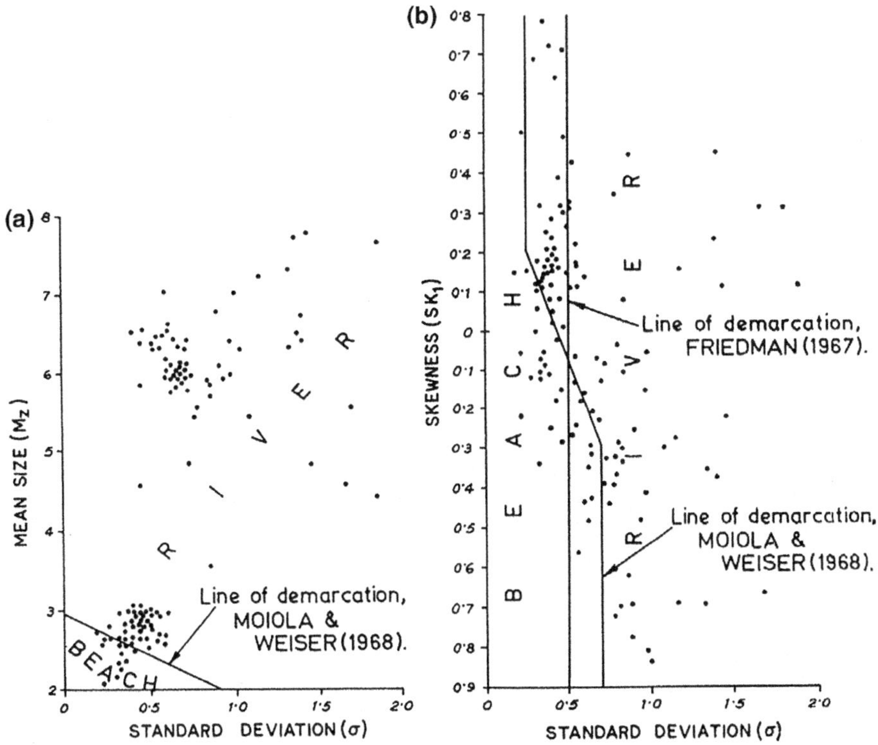

Fig. 6.16 Plot of Mean size (M_Z) versus Sorting (σ_1) and Sorting (σ_1) versus Skewness (SK_1) in the demarcated fields of river and beach environments

tidal creek without any freshwater discharge from the landward side, its grain size parameters excellently correspond to those of a river system. A few plots of samples from the swash bar, however fall within the beach environment (Fig. 6.16) and reflect similarities in hydrodynamic conditions within both of these two geomorphic domains.

6.9 Summary

There is rhythmicity in the nature of deposition, which perhaps indicates the depositional pulses for the ebb and flood flows through the rivers. The inflection points in the curves between successive ebb-flood cycles are marked at 2.5–3.0 phi, 3.5 phi, 5–6 phi and 8 phi levels respectively. Such rhythmicity in the nature of depositional behaviour of tidal sediments is supposed to be highly process responsive. Sandy samples (ranging from 1.5 to 4 phi) collected from the point bars and mid-channel bars, however, reveal a completely different pattern from that of sandy silty samples. The cumulative curves for these samples show a non-linear pattern and are nicely comparable to those of the other point-bar samples. These samples show two major inflections, one at 2.25 phi and the other at 3.0–3.5 phi. These inflections divide the curves into the three subpopulations of rolling, saltation and suspension respectively. The saltation population constitutes about 75 % of the total material, the rest being deposited by either rolling or suspension. These samples are collected from unidirectional flow areas like runnels or depressed channels within point bars and mid-channel bars. Silty sediments having range from 4 to 9 phi collected from similar areas also show exactly the same pattern wherein the three above-mentioned subpopulations can be well recognised. In these curves, the inflections take place at 6–7 phi respectively. The central saltation population constitutes about 70 % of the total. These sediments do reflect their deposition based on tractive movements of water in a unidirectional flow condition.

As a matter of fact, the grain size of the river sediments of Thakuran reflects the nature of source sediments and their hydrodynamic conditions of deposition. Generally, erosion prevails towards the seaward reach of the rivers with high wave energy, and deposition dominates in the landward reaches with a relatively quieter environment. Thus finer muddy sediments are deposited on the riverbanks, flanks of the mid channel bars and point bars of Thakuran River with low depositional energy.

References

Das GK (2015) Estuarine morphodynamics of the Sunderbans. Coastal research library, vol 11. Springer, Switzerland, 211p

DeMaeyer P, Wartel S (1988) Relation between superficial sediment grain size and morphological features of the coastal ridges off the Belgian coast. In: DeBoer PL et al (eds) Tide-influenced sedimentary environment and facies. D Reidel Publication Company, pp 91–100 (530p)

Evans G (1965) Intertidal flat sedimentation and their environment of deposition in the wash. Q J Geol Soc 121:209–241

Flemming BW (1999) Mass physical properties of muddy intertidal sediments: some applications, misapplications and non-applications. Intertidal mudflats; properties and processes. Plymouth, England

Folk RL (1954) Sedimentary rock nomenclature. J Geol 62:345–351

Folk RL, Ward W (1957) Brazos river bar-A study in the significance of grain size parameters. J Sediment Petrol 27:3–26

Friedman GM (1967) Dynamic processes and statistical parameters compared for size frequency distribution of beach and river sands. J Sediment Petrol 37:327–354

McLaren P (1981) An interpretation in grain size measurements. J Sediment Petrol 51:611–624

Middleton GV (1976) Hydraulic interpretation of sand size distributions. J Geol 84:405–426

Moiola RJ, Weiser D (1968) Textural parameters: an evaluation. J Sediment Petrol 38:45–53

Pejrup M (1988) The triangular diagram used for classification of estuarine sediments: a new approach. In: DeBoer PL, Van Gelder A, Nio SD (eds) Tide-influenced sedimentary environments and facies, D Reidel Publication Company. Holland, pp 289–300 (530p)

Rahmani RA (1988) Estuarine tidal channel and nearshore sedimentation of a late Cretaceous epicontinental sea, Drumheller, Alberta, Canada. In: de Boer PL, Van Gelder A, Nio SD (eds) Tide-influenced sedimentary environments and facies. D Reidel Publishing Company. Holland, pp 433–471 (530p)

Sagoe KMO, Visher GS (1977) Population breaks a grain size distribution of sand. A theoretical model. J Sediment Petrol 47:215–310

Shepard FP (1954) Nomenclature based on sand-silt-clay ratio. J Sediment Petrol 24:151–158

Spencer DW (1963) The interpretation of grain size distribution curves of clastic sediments. J Sediment Petrol 33:180–190

Tanner WF (1959) Sample components obtained by the method of differences. J Sediment Petrol 29:408–411

Visher GS (1969) Grainsize distribution and depositional processes. J Sediment Petrol 39:1074–1106

Chapter 7
Sedimentary Structures

Abstract The exposed portions of the Thakuran River bed and the intertidal mid-channel bars reveal various bedforms of tidal origin. Sand flats with small and large-scale bedforms dominate throughout the downstream portion of the tidal shoals, whereas mudflats possess fewer physical structures. The presence of megaripple marks and ripple marks as bedforms indicates that the river extends to the upper part of the lower flow regime. Both the surface and internal structures are indicative of tidal influence. Surface structures like small ripple marks, linguoid ripples, megaripples, sandwaves, rill marks, rhomboid marks etc. are recognised. Internal structures like flaser and lenticular type are highly indicative of flood and ebb cycles. Zones with distinctive bedform characteristics can be delineated based on their morphological variations and orientations. Moreover, 5–15 cm-wide intertidal belts are populated by hybrid bedforms. The structures indicating modification of bedforms due to the fluctuating velocity and depth during ebb and flood periods include: (i) skewed spurs, (ii) ripplefans, (iii) flat-crested megaripples and sandwaves, (iv) dissected channels and (v) microdeltas.

Keywords Bedforms · Ripples · Megaripples · Sand waves · Ripplefans · Rill marks · Rhomboid marks · Skewed spurs · Flaser bedding · Swash marks · Backwash marks · Thakuran river · Sunderbans

The Thakuran River channel in the Hugli-Matla estuarine complex of eastern India experiences a semi-diurnal, spring tidal range above 5 m near the mouth (Das 2015). The tidal prism covers the entire 80 km length of the river. Most of the sediments of the river are recycled by tidal currents due to the extreme paucity of freshwater discharge from the headland, except during the monsoonal months. The intertidal sand flats are ornamented by bedforms of various kinds and scales. Several scales of unsteadiness are marked in the tidal process. Among many such processes, the following are the most important: (i) fluctuations in water depth and velocity associated with ebb and flood periods, (ii) flow reversals associated with ebb and flood, (iii) neap- and spring-tidal flow variations, and (iv) seasonal tidal variations with equinoctial cycles. A single bedform type and the assemblage of

G.K. Das, *Tidal Sedimentation of the Sunderban's Thakuran Basin*,
DOI 10.1007/978-3-319-44191-7_7

bedform populations display features due to one or more of these scales of unsteadiness at a particular point of time. Thus bedforms in a tidal situation often exhibit a quasi-equilibrium form under the effects of unsteadiness.

With a purpose to examine the internal manifestations of bedforms generated of the mid-channel bars, several L-shaped trenches were dug during the time of emergence. Megaripples bedding, parallel stratification, horizontal stratification, reactivation surfaces in cross-bedded units and penecontemporaneous deformation structures have been encountered as internal structures.

7.1 Bedforms Characters

Field study for years revealed that the bedforms, though not static, are permanent features. An individual bedform zone is a reflex of the prevailing tidal domain in the area. The tidal domains have a general tendency to shift their positions and these, in turn, change the bedform characteristics of a particular locality (Das 2016). The bedform zones and the interzonal areas are thus dynamic features.

Inherent unsteadiness and reversals of tidal currents, as well as bedform-current interactions, also cause frequent changes in bedform architecture. Smaller bedforms quickly change their orientations, but large sandwaves do not, in response to flood-ebb change over (Dalrymple et al. 1978; Boothroyd 1978). The following deliberation presents an account of bedforms, generated in the meso-macrotidal Thakuran River, as comprehensively as possible.

7.2 Sedimentary Structures on the Mid-Channel Bars

Three different scales of bedforms are recognisable in the mid-channel bars of Pachim Sripatinagar of the Thakuran River, viz. the small-scale, the intermediate-scale and the large-scale, and these have been classified based on the scheme of Reineck and Singh (1980). The small-scale bedforms include all varieties of ripples, the intermediate-scale embodies two different megaripples types and the large-scale features comprise the sandwaves (Table 7.1).

7.2.1 Megaripples

7.2.1.1 Straight-Crested Megaripples

The surfaces of mid-channel bars are extensively sculptured by trains of straight-crested megaripples. These megaripples are two dimensional forms with low height (H) ranging between 0.15 and 0.43 m, moderate wavelength (L) ranging

Table 7.1 Bedforms of Thakuran and their dimensions

Bedform type	Length (m)	Height (m)	Span (m)	Steepness (L/H)
A. Small-scale				
Ripples n = 143	0.05–0.1	0.005–0.009	0.08–0.12	10–24
B. Intermediate-scale				
I. Straight-Crested Megaripples (= Megaripples Type-I of Dalrymple et al. 1978) n = 61	4.6–12.3	0.15–0.43	5.9–13.7	14.8–57.14
II. Undulatory Megaripples (= Catenary and Sinuous Megaripples of Allen 1968) n = 34	2.43–4.88	0.14–0.32	7.3–11.58	9.48–29.16
C. Large-scale				
Sandwaves n = 35	15–25	0.65–1.4	43–97	15.38–27.54

from 4.6 to 12.3 m (Fig. 7.1e) and a long span ranging from 5.9 to 13.7 m (Table 7.1). They are parallel to each other over considerable distances but exhibit minor sinuosity (Fig. 7.2). These megaripples show broadly convex-up profiles and their crestal heights remain more or less constant without showing any well-developed scour pits in the troughs. The troughs contain small ripples with crestal orientations perpendicular to the megaripples crests. This is because of the generation of a secondary-flow pattern along the troughs of megaripples during falling of the water level.

The L/H ratio of the megaripples ranges from 14.87 to 57.14. Wavelength versus height maintains a definite correlation (Fig. 7.3a). The graphic mean size (M_Z) of these megaripples ranges from 2.45 to 2.96 phi with a mean of 2.66. Trench sections on mid-channel bars often reveal intact preservation of crests and troughs of megaripples (Figs. 7.4 and 7.5).

The lee slope measurement of the megaripples near the crests was around 32°, whereas, near the base, the angle was reduced to around 10°. However, in areas where these bedforms are modified by a change in flow direction with falling water level, several spurs emerge from the slip faces, which are further mantled by ebb-oriented, falling-stage ripple-fans (Figs. 7.6, 7.7 and 7.8). The ripplefans are modification structures in front of slip faces and are formed by the formation of helical flow-cells in between the spurs (Allen 1984).

These megaripples are formed at somewhat lower velocities (max. velocity of 70 cm/s and a water depth of 2 m) than that required for the undulatory or lunate megaripples (Dalrymple et al. 1978; Reineck and Singh 1980). In some instances, small-scale wave ripples climb over the megaripples or interfere with their crestal trains (Fig. 7.9).

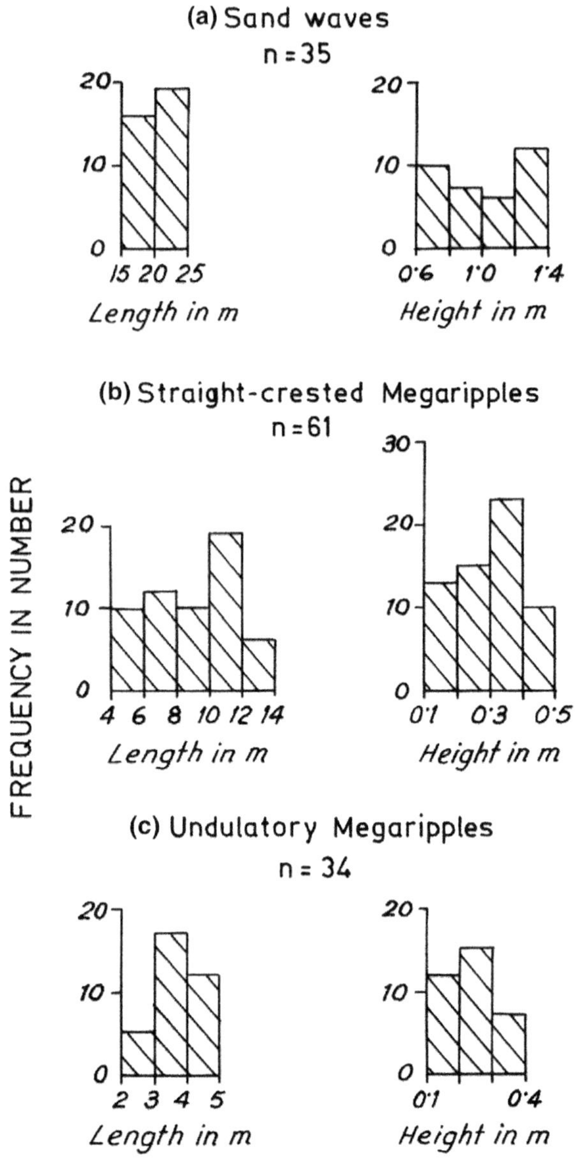

Fig. 7.1 Histograms of distribution of length and height values of sandwaves, straight-crested megaripples and undulatory megaripples

7.2.1.2 Undulatory Megaripples

They possess long wavy or undulating crests and are devoid of well-developed scour pits in front of the slip face. In this respect they differ from lunate megaripples of Reineck and Singh (1980) or Type-II megaripples of Dalrymple et al. (1978) in which the crest line is broken and megaripples possess distinct scour pits in front.

Fig. 7.2 Typical appearance of ebb-oriented, straight-crested megaripples (Type-I) with superimposed linguoid ripples on both stoss and lee slopes

Both in-phase and out-of-phase arrangement of undulations are present in a single megaripple train. The crests show forward tongue-like projections and steep profiles in contrast to those of the straight-crested megaripples. The megaripple surface is ornamented by linguoid ripple and these ripples are produced during emergence. Allen's (1968) catenary megaripple corresponds to this type.

They form at higher velocities (>70 cm/s as specified by Reineck and Singh, 1980) than that of the straight-crested type. The wave-lengths (L) of the undulatory megaripples ranges from 2.43 to 4.88 m, heights (H) from 0.14 to 0.32 m (Fig. 7.1f) and spans from 7.3 to 11.58 m (Table 7.1). L/H ratio ranges from 9.48 to 29.16 and so less than that of straight-crested megaripples. A definite correlation also exists between length and height for these megaripples (Fig. 7.3b). Graphic mean size (Mz) of sediments constituting these megaripples ranges from 2.31 to 2.86 phi with mean of 2.42. The undulatory megaripple is supposed to be an intermediate form between the straight-crested megaripples of Reineck and Singh (1980) or type-I megaripples of Dalrymple et al. (1978) and lunate megaripple of Reineck and Singh (1980) or Type-II megaripple of Dalrymple et al. (1978).

7.2.2 Sandwaves

They are the largest scale bedforms observed on the mid-channel bar. They are flood-oriented, two-dimensional forms and appear mostly on the floors of the flood-dominated portions of bar surfaces. They lack scour pits in their troughs and

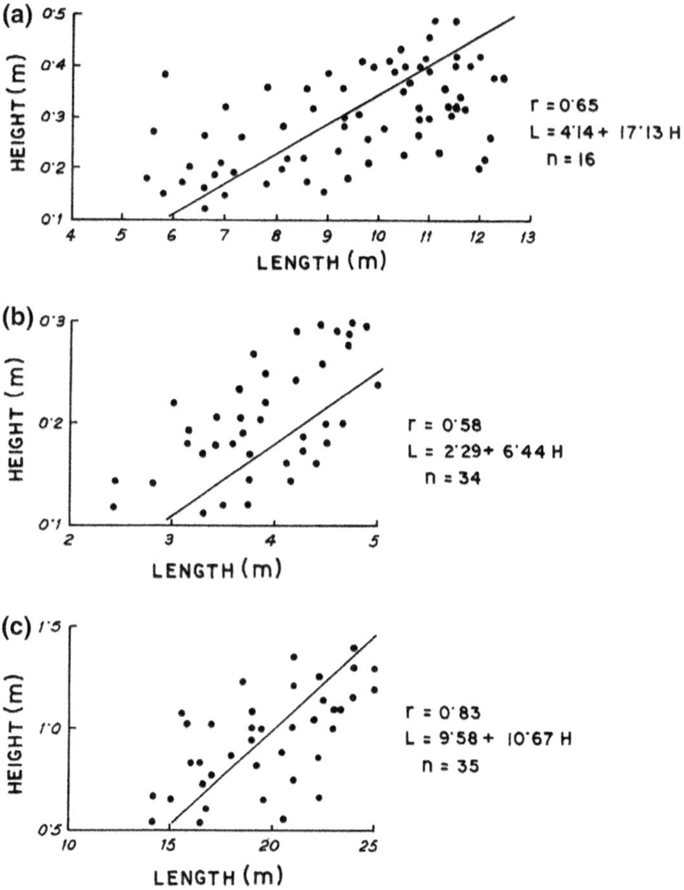

Fig. 7.3 Plot of length versus height of straight-crested megaripples (**a**), undulatory megaripples (**b**) and sandwaves (**c**)

spurs on their slip faces. Ripples superimposed on sandwaves have been partially planed off by the preceding ebb-flow. Their wavelengths range from 15 to 25 m, heights from 0.65 to 1.4 m and spans 43 to 97 m (Table 7.1). Their crests are straight to sinuous (Fig. 7.10). Crestal heights remain constant along span. Plotting of wavelength versus height of sandwaves reveals a strong correlation (Fig. 7.3c). The length-height ratio ranges from 15.38 to 27.54.

These sandwaves are similar to the "rippled sandwaves" (Dalrymple et al. 1975, 1978; Elliott and Gardiner 1981) in terms of their scale of formation and superimposition by small-scale ripples. The relatively larger set of 'megarippled sandwaves' described by (1978, 1990) is not found in the current study area. Despite a general similarity in the appearance of these rippled sandwaves with those of the Bay of Fundy and the Loughour estuary, they show marked differences in other attributes (Table 7.2).

Fig. 7.4 Section parallel to river length through mid channel bar sand flat reveals megaripple phase followed upward by a plane bed sequence with a plane of truncation in between. The internal laminations of the megaripple phase show wavy undulation due to penecontemporaneous deformation. Megaripple crest with mud couplets and tidal bundles with sand-mud layers are also seen

Fig. 7.5 Section through Dhanchi mid channel bar showing tidal beddings with alternation of sand mud laminae. Note the deformed foresets in the lower unit. The surface shows current crescents. Rular is 30 cm

7.2.3 *Internal Physical Structures of the Mid-Channel Bars*

With a purpose to examine the internal manifestations of the bedforms generated on the mid-channel bars, several L-shaped trenches were dug during the time of emergence. Excepting g a few mid-channel bars (e.g. that of Paschim Sripatinagar, Dhanchi), many of the mid-channel shoals undergo complete submergence during

Fig. 7.6 Ebb-oriented straight-crested megaripples with sharp-crested spurs attached to the slip face at an angle of 45°–80° and mantled by ebb-oriented falling-stage ripplefans. The main current is from *upper left* to *lower right*. Scale is 30 cm

Fig. 7.7 Ebb-oriented straight-crested megaripples in decaying stage. Slip face of megaripples not clearly discernable in places. Note the continuation of the crests of small ripples from megaripple surface to the crests of spurs up to underwater ripples. Ripple fanning is not prominent in front of slip face

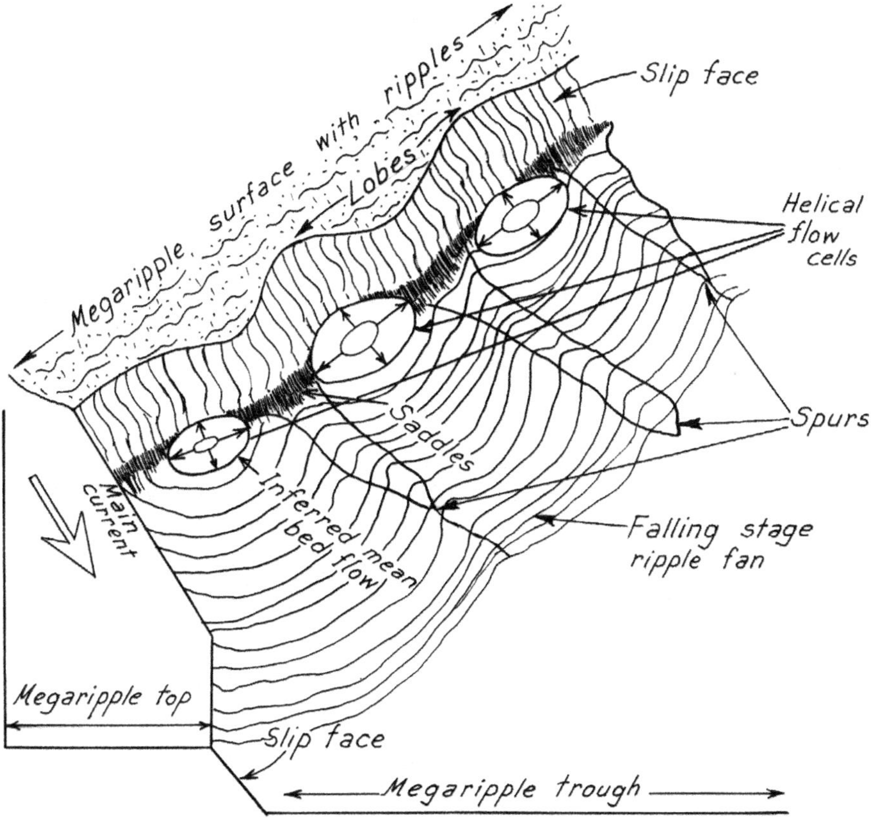

Fig. 7.8 Ripple fan development in the separated flows to lee (*slip face*) of straight-crested megaripples (modified after Allen 1984)

both flood- and ebb-tides. As a result, there was limited opportunity for examining the trench sections. However, the following types of internal structures were identified and examined in detail.

7.2.3.1 Megaripple Bedding

The upper surface of this type of bedding is an undulatory, non-erosional surface of the megaripples preserved as foreset. The megaripple form is capped by a thin veneer of mud with thicknesses of a few mm. The megaripples in section are 50–60 cm long, 12–18 cm high and correspond well with that of the surface megaripples.

This bedding type is the internal manifestation of megaripples that migrate with flood currents. The thin mud veneer at the crests and troughs of megaripples

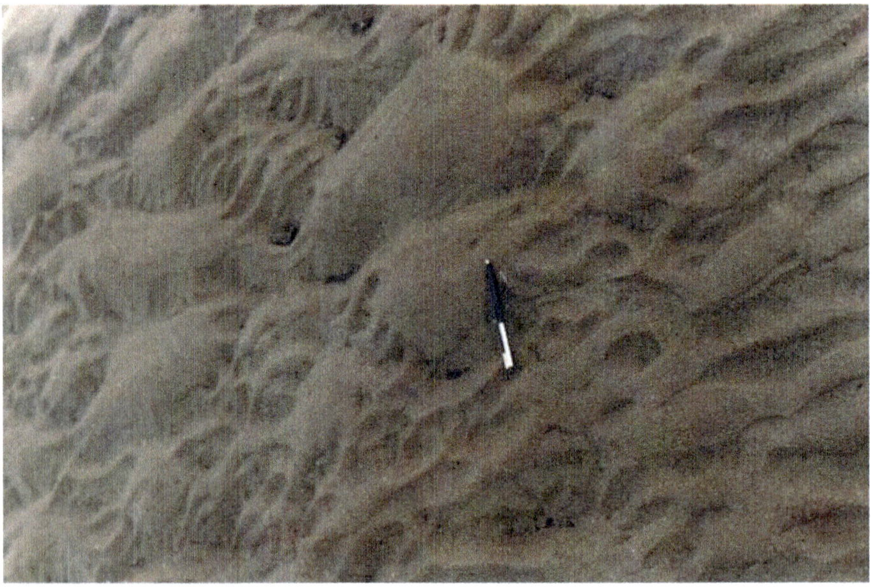

Fig. 7.9 Small-scale wave ripples climb over the megaripples and interfere with the crestal trains of megaripples. Flattening of crests of megaripples often obliterate the superimposed ripples. Pen is 15 cm

Fig. 7.10 Small-scale, ebb-oriented ripple aprons on the crest of a flood-oriented sandwave. Note also the channel system extending along the trough line of the sandwave. Superimposed ripples show fanning on the lee face of the sandwaves

Table 7.2 A comparison of the rippled sand waves of the present study area, Bay of Fundy and Loughour estuary

Parameters	Bay of Fundy Dalrymple et al. (1978)	Loughour Estuary Elliott and Gardiner (1981)	Thakuran River (Present Worker)
Orientation	Not stated	Majority flood oriented	Flood-oriented
Lee-slope angle	Much below angle of repose, 10–20°	Steep angle of repose, 26–30°	Steep angle of repose, 25–28° (Comparable to the megaripples lee slope)
Nature of dominant and subordinate tides	Dominant and subordinate tides not differentiated	Dominant and subordinate tides differentiated	Dominant and subordinate tides well-differentiated
Grain size	Coarser than medium sand (Av. Mean size = 1.44 phi)	Fine sand	Medium to fine sand (Av. Mean size = 2.66 phi)
Stability	Do not form stable bedform field; shows disequilibrium and decaying form of megarippled sandwaves	Stable bedform field; without showing progradation or retrogration to another bedform type	Stable bedform field; do not prograde or retrograde to other bedform type
Tidal amplitude	Semidiurnal Tide > 16 m	Semidiurnal Tide > 7 m mean neap: 4.0 m mean Spring: 8.1 m	Semidiurnal Tide > 4.5 m mean neap: 3.0 m mean Spring: 5.5 m

presumably indicates deposition of mud during periods of slackening of tidal currents. The angle of repose of lee slopes of surficial megaripples corresponds well to the angle of inclination of the foreset laminae of subsurface cross-bedded units.

7.2.3.2 Parallel Stratification

The megaripple bedding in most cases is followed upward by parallel stratification (Harms and Fahenstock 1965) with distinct tidal bundles characterised by alternations of fine sand-mud couplets (Fig. 7.4). The thickness of individual layers range in mm-scale, being confined within the cm-scale sets of parallel stratification.

The fine sand and mud are quite obviously deposited from suspension by slowly moving tidal currents. The intercalated mud stringers within fine sandy laminations also imply deposition from tidal slackening. Parallel stratifications have non-erosional contact with the underlying unit.

7.2.3.3 Horizontal Stratification

Trench sections on the sandy flat surfaces of mid-channel bars exhibit sets of horizontal stratifications as internal structure. The laminae within the sets of horizontal stratifications are often ill-defined or faintly marked (Fig. 7.5). Some laminae become quite prominent due to concentrations of heavy minerals. The sets of horizontal stratifications range from less than a cm to a few cms. The lower bounding surface of each set is erosional, horizontal to subhorizontal and planar. The horizontal stratifications are essentially made up of sand-sized particles, and in this respect they differ from parallel stratifications which are made up of finer sized particles and essentially maintain a non-erosional contact with the underlying unit.

Horizontal stratification is the internal manifestation of plane beds or featureless surfaces of sand transport (Harms and Fahenstock 1965). Such surfaces generally occupy the relatively more elevated areas of mid-channel bars where sand moves in a streaming fashion without any turbulence.

7.2.3.4 Reactivation Surfaces in Cross-Bedded Units

On some occasions, convex-upward reactivation surfaces are present within cross-bedded units (Fig. 7.4). Mud drapes are absent in the cross-bedded units (Houthuys and Gullentops 1988). Reactivation surfaces indicate pause planes (Boersma and Terwindt 1981) within the cross-bedded units and appear as discontinuities in the patterns and attitudes of foreset laminae. Fluctuations or changes in flow mechanisms or direction in tidal currents is responsible for the generation of such discontinuity or pause planes.

7.2.3.5 Penecontemporaneous Deformation Structures

These structures include some disturbed, distorted or deformed sedimentary layers (Fig. 7.4) in the mid-channel bar deposits. The deformations occur simultaneously with or shortly after the deposition of sediments. The deformed laminae remain confined within a single unit within a sequence of undeformed units.

In some places, the deformations are marked by slightly disturbed layers with minor undulations or folding. The deformed layers (2–3 cm thick) are confined within absolutely undeformed horizontal stratifications. The deformations broad warps of laminae at a depth of 3–4 cm from the mid-channel bar surface.

Deformations noticed in other trench sections are much more complicated, showing convolutions of laminae within undeformed overlying and underlying units (Fig. 7.4). The deformations are confined within 3–4 cm thick unit. Very fine alternations of sand and mud laminae reflect a tidal-bedding origin of the deposits.

Both varieties of deformation in the study point to an origin from plastic flow of materials resultant from unequal shear as explained by McKee et al. (1962a).

7.2.4 *Modification-Features Resultant from Unsteady Flow in Tidal Environment*

Tidal processes are characterised by many scales of unsteadiness in normal natural settings. Of the many such causes of unsteady flow, the following seem to be the most important:

(i) Fluctuating velocity and depth within ebb and flood periods
(ii) Flow reversals associated with successive ebb and flood periods
(iii) Contrasts in neap and spring tidal flow
(iv) Seasonal variation in tides between equinox and solstice, i.e. with equinoctial cycles.

Although bedforms are generally considered to be the result of a principal flow and related sediment-transport conditions, several effects of unsteadiness in flows are not commonly emphasised. But individual bedforms and the overall bedform assemblages at a point of time do reveal features due to one or more of the above mentioned scales of unsteadiness.

7.2.4.1 Modification Bedforms Resultant from Fluctuating Velocities and Depths with Ebb and Flood Periods

For large-scale bedforms like megaripples and sand waves, modification takes place when the flow velocity falls during the waning phase of ebb or flood currents from peak-flow conditions. This velocity is unable to migrate the slip faces of these large-scale features. Waning of flood periods involve a progressive reduction in flow velocity accompanied by an increase in flow depths. On the contrary, during waning ebb periods, a decrease in flow velocities is accompanied by a decrease in flow depths. As a consequence, the intertidal mid-channel-bar surface undergoes a brief period of high flow velocities with decreasing depth immediately prior to emergence. These changes in flow conditions are referred to as "falling stage" (Elliott and Gardiner 1981), and, during this state of change of ebb periods, the modification features described in the following sections have been noticed. Observations of these features on the bar surfaces can only be made during a short period after the ebb flow.

(i) **Skewed Spurs**

These features appear under direct observations and measurements during the short spans of time limited within low-water periods after the recession of tides. Some of the straight-crested megaripples are unique by possessing sharp-crested, skewed spurs which project down-current from the slip face at an angle of 60 to 90° (Figs. 7.6 and 7.7). The slip face is plannar or curved. The portions of the lee slope that projects down-current are called the "saddles", whereas, the reentrent sections form the "lobes" (Fig. 7.8). The spurs in the current study area are generally attached to the saddles of the slip faces (Fig. 7.8).

The spacing between two successive spurs ranges from 0.85 to 1 m and the spurs continue for 1–3 m down-current before dying out (Figs. 7.6 and 7.7). The elongated scour pits that occur in between the spurs contain ripplefans, which run continuously across the spurs. The ripple trains surmount the crests of the spurs in almost all cases. The ripplefans, however, are not always associated with spurs and grooves. Such features known as "ripple scours" (Fig. 7.6) have been described by Potter and Pettijohn (1963) and Allen (1984). Straight-crested megaripples with skewed spurs are more-or-less permanent features of mid-channel-bar surfaces, although they shift their position with time.

Spurs attached to megaripples have been described by Allen (1968), Knight (1972), Dalrymple et al. (1978) and Elliott and Gardiner (1981). In Allen's experiment, these scours and spurs were developed by pairs of contra-rotating vortices with helical-spiral motions. Knight (1972) described the skewed vortices to be the product of a change in flow-direction during the falling stage of ebb flow. Elliott and Gardiner (1981) supported the interpretation of Knight (1972). The present researcher, in accordance with the interpretation of Elliott and Gardiner (1981) holds that with the fall of water level from the earlier deeper flows, the flow direction changes with increases in local slope in front of the slip face. Vortices developed in front of the slip face produce evenly-spaced swept areas and spurs. With further fall of water levels, the ripplefans superimpose the spurs and grooves (Fig. 7.8).

(ii) **Ripplefans**

The ripplefans (Allen 1968) are mostly superimposed on the stoss slope of sandwaves and on the troughs of the straight-crested megaripples (Figs. 7.6 and 7.7). However, their occurrence along the troughs of straight-crested megaripples is by far the most common. They have an overall fan shape with the ornamentation of an ordered pattern of asymmetrical ripples. Their continuous crestlines fan away from the slip face of megaripples (Figs. 7.6 and 7.7). Wavelengths of these ripplefans break down and change into linguoid ripples as they ascend the stoss slopes of the next succeeding megaripple.

In the present field of study, the ripplefans develop at the falling stage conditions when a number of helical flow cells are generated in front of the slip face (Fig. 7.8). These "cells" are divided transversely along imaginary stream lines called "seams" (spurs) towards which the ripples have a facing component (Figs. 7.6 and 7.7). The seams usually coincide with the crest of spurs (Allen 1984). This is substantiated by the fact that the point of origin of the ripplefan is located close to the foot of the slip face. This suggests that the separation of eddies has effectively contracted with distance. Further, the skewed nature of the ripplefans down-slope points to increasing influence of the slope as the water level falls (Elliott and Gardiner 1981).

(iii) **Planned-off Crest of Megaripples and Sandwaves**

At places, the megaripples and sandwaves are planed off removing the superimposed ripples, and the bedform crests become wider. The superimposed ripples may be retained only in isolated patches on the generally planed-off surface. These surfaces originate just before the bedform emergence when the plane bed flow regime prevails only for a short while.

The superimposed ripples formed earlier are largely obliterated due to swift erosion of the large-scale bedform crests. These modulations are obvious responses of bedforms to changing flow velocities and depths within flood-ebb periodicity.

Boothroyd and Hubbard (1975), Dalrymple et al. (1978, 1990), and Elliott and Gardiner (1981) described these structures from intertidal sand bodies under meso- and macrotidal settings. Boothroyd and Hubbard (1975) considered them to constitute a separate class of bedforms having a specific position in a flow-regime scheme, rather than as results of modification of pre-existing bedforms as suggested by Dalrymple et al. (1978).

(iv) **Dissection Channels and Microdelta**

The spurs attached to the slip face of straight-crested megaripples are occasionally cut by narrow steep sided channels transverse to the spurs. Sediments eroded by these channels are deposited at the mouth of the channels as microdeltas that grade towards the troughs of the megaripples. In plan, the microdelta is typically triangular with a lobate front. The surface of the microdelta is often characterized by linguoid ripples developed during late-stage run off.

(v) **Strandlines**

Slip faces of megaripples and sandwaves are very commonly etched or marked by closely spaced parallel lines that reflect slight reworking of the slip face by successively lower water levels (de Klein 1970; Reineck and Singh 1972). With falling water, a series of such markings are formed at different levels. They indicate sinking water levels in an area exposed to intermittent subaerial exposure.

7.2.4.2 Modification Bedforms Due to Flow Reversals

The flood-oriented megaripples and sandwaves on the mid-channel bars are exposed in many areas following the ebb tide. Observations in the field reveal that the crests of these large-scale structures are occasionally rounded and their upper surface supports small, asymmetrical, ebb-oriented ripples. The subordinate ebb tide has limited reworking ability to change wholesale the orientation of the large-scale structures. The ebb-oriented ripples also occur locally as low-angle mounds. Heights of these ripples range from 2 to 8 cm, whereas, their slip faces range from 12 to 20°.

7.2.4.3 Modification Bedforms Due to Contrasts Between Neap and Spring Conditions

The megaripple fields in the study area are observed generally shift their positions in response to contrasted neap-spring cycles. Zones of decaying megaripples appear as the most conspicuous feature formed under these conditions.

(i) **Decaying Megaripples**

Megaripples often retain their orientations during neap-spring cycles. They maintain their prominence both in conditions of waxing and waning of neap-spring cycles. Under favourable conditions, the undulatory megaripples and megaripples of Type-1 increase in wavelength, diminish in height and therefore become less distinct and less prominent (Elliott and Gardiner 1981). Occasionally, sediment transport along the ripple fields enhances decaying of megaripples. Thus megaripples of low height are produced by transformation, and these megaripples are often superimposed by linguoid ripples (Das 2016). The slip faces are generally weakly discernable or may be absent in extreme cases.

7.2.5 *Description of Small-Scale Bedforms*

7.2.5.1 Ripples

Ripples are the smallest bedforms on sand flats. There are many kinds of ripples of which the complex interfering type is quite abundant. Each type of ripple extends over its own separate field covering large areas, particularly on the topographically higher levels of mid-channel bars. On the contrary, in topographically lower area, different kinds of ripples occur in close association with one another.

Generally the ebb-oriented and flood-oriented ripples are asymmetrical and linguoid in form. They occur in exclusive fields of their own. Both symmetrical and asymmetrical straight-crested ripples of short wavelength are found locally in areas of wind-driven waves active in shallow water. Areas of interference of waves and currents produce interference ripples of very complex morphology. Ripples of more than one type occur in close association where the hydrodynamic conditions change within small areas. The various kinds of ripples are described in the following paragraphs.

(i) **Straight-Crested Ripples**

These ripples appear as both symmetric and asymmetric forms. They have more-or-less straight crests. Crestal trains of successive ripples run parallel for a few metres. The surface undulation is minor because of low ripple heights ranging from 0.25 to 0.8 cm (Fig. 7.11). Wavelength is short and ranges from 5 to 11 cm (Fig. 7.12a). A good linear correlation exists between ripple wavelength and height (Fig. 7.13b).

The straight-crested ripples are formed by local, short-period, wind-driven waves that are formed in shallow waters. Thus, their origin is related to periods of moderate-to-strong winds active on sand flats.

(ii) **Linguoid Ripples**

These ripples are crescentic in plan. Their crests are arcuate or tongue-like having down-current closures (Fig. 7.2). The ripples produce appreciable surface undulations as heights range from 1 to 3.4 cm (Fig. 7.12c). Wavelengths range from 8 to

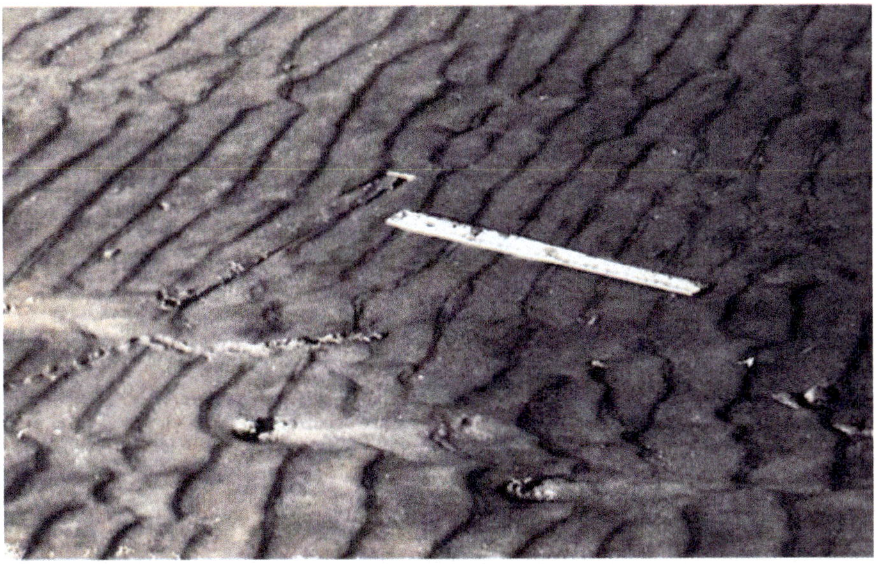

Fig. 7.11 Wave ripples grade to straight-crested ripples. Current crescents are opposite to ripple lee-directions and indicate reversing of tidal currents. Ruler is 30 cm

19 cm. Plots of wavelength versus height reveal a nicely positive linear correlation (Fig. 7.13a).

Linguoid ripples commonly occupy creek floors on mid-channel bars where the water depth is around a metre, particularly during the flooding time. They are generally oriented parallel to flood flow but may also occur in an ebb-oriented way. These ripples display definite field evidence of crestal reversal with tides. Linguoid ripples laterally grade to small wave ripples or to straight-crested ripples with decreasing water depth.

(iii) **Small Wave Ripples**

These ripples are asymmetric in plan and exhibit crestal sinuosity. They show close association with straight-crested ripples. They have a steep lee-slope and a very gentle stoss slope (Fig. 7.14). Wavelengths range from 7.0 to 8.5 cm and height from 0.08 to 0.11 cm. A positive linear correlation exists between their wavelength and height.

Small wave ripples occur as strike-wise variations from straight-crested ripples and reflect a shallow-water origin on the mid-channel-bar surface. These ripples are also formed by wind-driven waves generated in shallow waters.

7.2.6 Some Characteristics Ripple Types and Other Structures of Tidal Origin

Although not exclusive to intertidal flats, some ripple types, viz. flat-topped ripples, ladder-back ripples and double-crested ripples, are very characteristic of intertidal

Fig. 7.12 Histograms of distribution of length and height values of various ripples

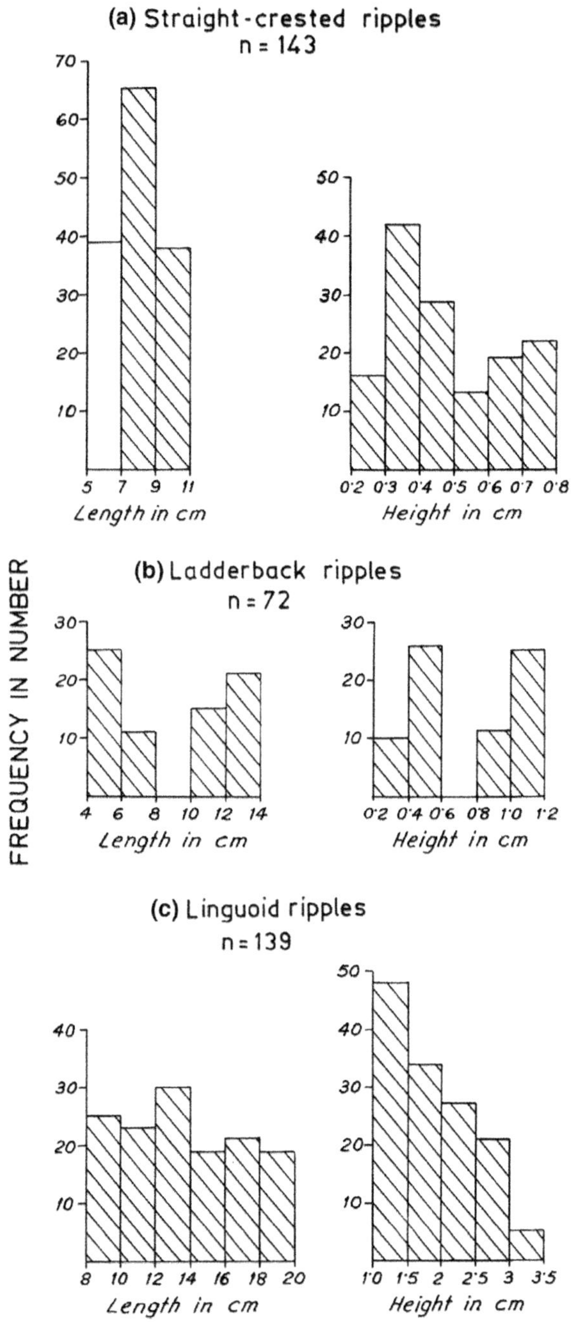

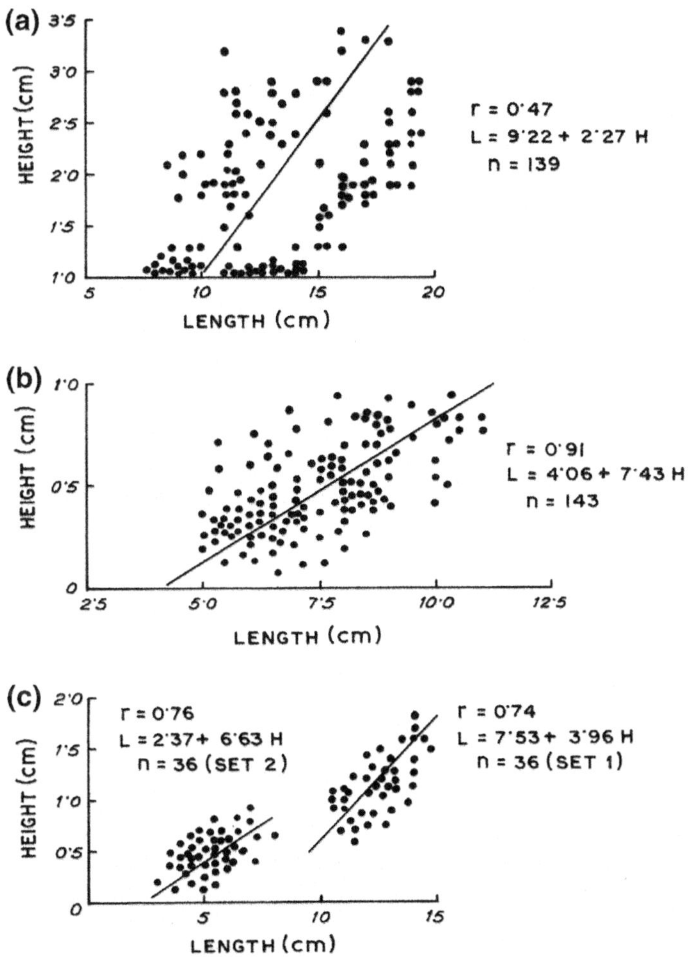

Fig. 7.13 Plot of length versus height of Linguoid ripples (**a**), straight-crested ripples (**b**) and Ladder-back ripples (**c**)

zones (Terwindt 1988) and have been studied based on the mid-channel bars of the Thakuran River.

7.2.6.1 Flat-Topped or Capped-Off Ripples

They occur superimposed on sandwaves and on crests of megaripples. The flattening of the crests is generally prevalent in linguoid ripples and asymmetric wave ripples. In flat-topped ripples, the ripple troughs are very narrow compared to the width of the crests (Fig. 7.15). The magnitude of the crestal flattening increases with increasing slope of the intertidal flat as the ripples approach the megaripple-trough

Fig. 7.14 Wave ripples on intertidal sand flat at Dhanchi. Ebb flow is towards left of the photograph. Note the flattening of the ripple crests and narrowing of ripple troughs. Upright trunks of mangroves are seen. Scale is 30 cm

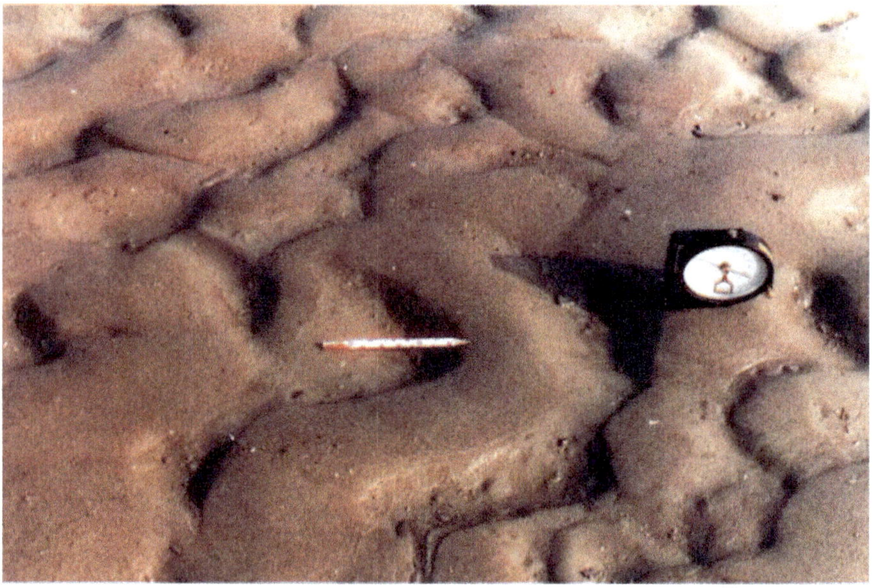

Fig. 7.15 Flat-topped linguoid ripples. Note the narrowing of the trough compared to width. Pencil is 12 cm

Fig. 7.16 Capped-off ripples. Wave ripples are modified to linguoid current ripples by the thin film of water during subaerial emergence. Sudden truncation of the ripple trains (marked by clinometer) is due to shear drag of surface ripples during emergence

margins. In extreme cases, the ripple crests are modified to more-or-less flat surfaces and maintain very little space for the ripple troughs.

Sometimes, the wave ripples with sinuous crests are seen to change into linguoid forms during the final draw down of water. At places, the cress of a superimposed younger set of interference ripples are the only ones flattened, leaving the crestal trains of the earlier set almost intact. The narrow elongated troughs of ripples are ornamented with the earlier set of ripple crests, obliquely or at right angles to the elongation direction of ripple troughs.

The mechanism of formation of flat-topped ripples is attributed to erosion (Fig. 7.16) induced by the enhanced shear of falling water (Sanders 1963; Bogardi 1961) or due to backwash at late-stage emergence (Klein de Klein 1970). According to Reineck and Singh (1980), the flattening is caused by 'capillary waves' that are produced by strong winds on the water surface. Flat-topped ripples are typical, though not exclusive, to tidal deposits (Terwindt 1988).

7.2.6.2 Ripples with Double Crests

Some domains of ripples in the intertidal flat display ripple with double crest lines having spacing about 5–6 mm apart. The ripples are generally asymmetric in profile and have almost straight crests, which often terminate laterally against the linguoid ripples. Double-crested ripples may form under the following conditions: (i) in stagnant pools of very shallow water by wind-generated waves and (ii) in ripple flats under wave-cum-current flow. The double-crested ripples most commonly occur in intertidal zones (Klein 1970; Reineck and Singh 1980; Terwindt 1988). Yet, they are not exclusive to such environment.

7.2.6.3 Ladder-Back Ripples

These are interference ripples in which two sets of superimposed ripples, which are oriented essentially at right angles to each other (Davis 1983). On most occasions, there is a difference in the scale of formations of the two sets, and the smaller ripples run along the troughs of larger ripples. In this field, many different configurations for these ripples exist varying from simple ladder-back forms (Fig. 7.17) to complex network of superimposed ripple trains. The superimposed, small current ripples often occur on straight or sinuous-crested ripples having crestal trends at almost right angles to each other.

Wavelengths of the larger set of ladder-back ripples range from 10 to 14 cm and that of the smaller set from 4 to 8 cm. The corresponding heights of ripples range from 0.8 to 1.2 and 0.2 to 0.6 cm (Fig. 7.12b). The plots of wavelength versus height of these two sets of ripples reveal a strong positive linear correlation (Fig. 7.13c).

Fig. 7.17 Ladder-back ripples. An older set of ripples is preserved in the trough of a younger set formed at right angles. The younger set exhibit multiple crests. Pen is 15 cm

Reduction of water depth, particularly during the ebb phase, controls the size and orientation of current ripples. With the decline of water depth, the size of current ripples decreases, resulting in superimposition of smaller ripples on the larger sets. On many occasions in the case described here, the smaller ripples run along the troughs of the larger ripples, because the water level drops below the crestal heights of the larger ripples. In addition, ladder-back ripples of other kinds originate through interference between two simultaneously-operating wind-induced waves.

The lowering of water depth also produces double crests in the larger set ripples of the ladder-back variety. The secondarily-formed parallel ripples occur along the troughs of the earlier larger set. Tidal-current bottom shear drives the initial current ripples until the water level drops to a critical depth where ripple migration ceases. Yet the shear stress on the troughs, where water depth is slightly higher, may still be capable to move sand grains and create secondary ripples of smaller amplitudes (Klein 1970).

7.2.6.4 Relics of Ripple Crests

The wave ripples in the intertidal flats are often modified by current during a falling water stage. If some water is retained in the troughs at the time of subaerial exposure, the lee faces of the ripples suffer partial destruction by wind-driven water and leave only the relics of ripples crests. The wind-driven waves often create small wave ripples along troughs of the earlier ripple set (Fig. 7.18). Reineck and Singh (1980) have described similar relics of ripple crests on the tidal flats of the North Sea.

7.2.6.5 Water-Level Marks on Ripple Surfaces

Water level marks also known as etch marks (Klein 1970) are supposed to be the modification structures characteristic of an intertidal sand-flat environment. Such etch marks form during a punctuated fall in water levels on emerging mid-channel bars. Etch marks occur on both megaripple and ripple surfaces (Fig. 7.19).

Following the suggestion of Klein (1970), the mechanics of water-level marks can be explained as follows. As the tide ebbs and progressively exposes various parts of the mid-channel bar, different scales of pools appear in the troughs of ripples and megaripples. Low velocity open-channel flow also occurs in troughs of larger bedforms. A slight increase in surface-wind velocity forms a small-wave pulse on the pool surface. This wave pulse laps against the slip face of the bedforms (megaripples and ripples) and erode parts of the bedforms at that water level. The gradual fall in water level coupled with repeated microwave pulses produce the etch marks. Continuation of this process produces a series of parallel water-recession marks.

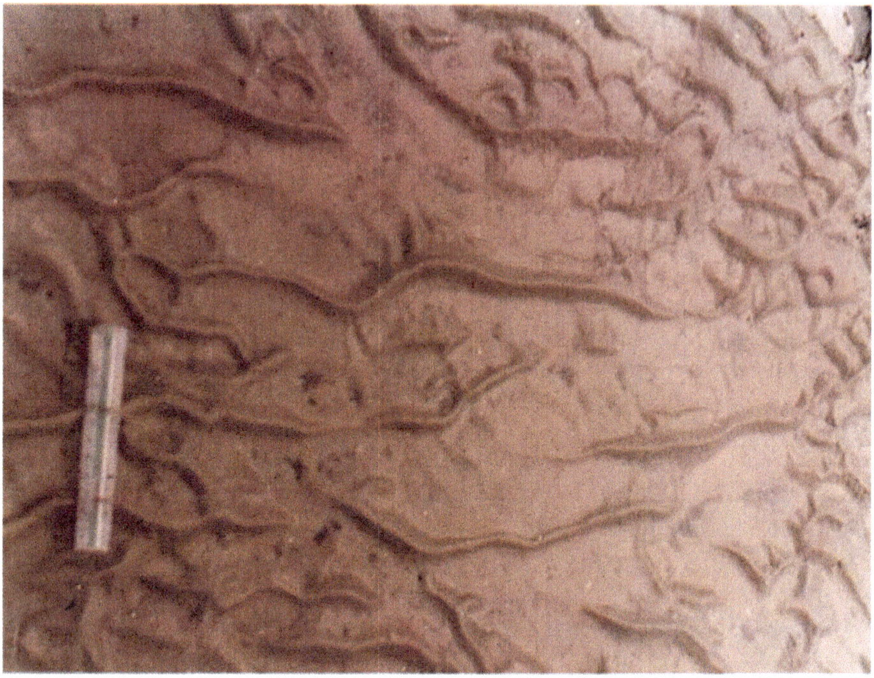

Fig. 7.18 Relics of the ripple crests. Small-scale wave ripples develop in the troughs. The relatively larger-ripples are destroyed by wind-driven current. Scale is 15 cm

Fig. 7.19 Water level marks on ripple slopes formed during irregularly (in phases) falling water level. Pen is 15 cm

7.3 Sedimentary Structures of the Point Bars

7.3.1 Description of Bedforms

7.3.1.1 Reversing Ripple

The point bar surface is often eroded by incoming wave attack so that the surface of the bar is lowered by 1.5–2 cm. Eventually, the upper sandy substratum is thrown into ripples by the reversing currents over the point bar. The ripples are generally asymmetrical, wave-cum-current, combined-flow ripples with rounded crests.

7.3.1.2 Longitudinal Ripples

These ripples named by Van Straaten (1961) are generally straight-crested without bifurcation. The ripple crests run more or less parallel to the current, and the wave propagation is at right angles to the current direction. Although wave action is supposed to be the major force for their formation, a simultaneous weak current flowing parallel to the crests also controls their shape (Reineck and Singh 1980).

The longitudinal ripples dominate on muddy, stable substratum. These ripples are generally symmetrical in form; their lengths range from 0.5 to 2.5 cm. The ripple crests continue uninterrupted for 10–100 cm. Reineck (1972) described longitudinal ripples as an erosive form in mud with about 40 % sand. Longitudinal ripples are characteristic of a very shallow water environment and generally form during falling water stages with eventual emergence.

7.3.1.3 Symmetrical Wave Ripples

These ripples have more or less symmetrical profiles and straight crests. Instead of having pointed crests and rounded troughs typical for wave ripples, these ripples possess rounded crests. The rounding of crests is thought to be a modification feature and results due to reworking of wave ripples during the process of emergence. Wavelengths of these ripples range from 5 to 6 cm and heights from 0.5 to 0.6 cm.

In many places, thin veins of sand (2 to 5 mm wide and a few cm long) occupy the narrow ripple troughs. Very commonly, the veins also show variations of width (Fig. 7.20). Thus, the veins appear as longitudinal bedforms and are good indicators of subaerial emergence or quasi-subaerial emergence of the ripple surface.

In sectional manifestations, these ripples show unidirectional foreset laminae. The forward motion of the wave somewhat stronger than the backward motion produces foreset laminae (Reineck and Singh 1980). The backward motion of a wave is strong enough to produce the symmetry of the ripples but unable to produce foreset laminae. The rounding of ripple crests takes place during the phase of ripple emergence. The sand veins perhaps indicate accumulation of transported sand along

Fig. 7.20 Wave ripples modified by swash. Crests of ripples are rounded. Thin veins of sand occupy the narrow ripple troughs. Note variations in width of the veins. Scale is 30 cm

the ripple troughs during subaerial exposures of the ripple surface. Their longitudinal alignment is chiefly controlled by the transportation of sand across the direction of wave movement.

7.3.2 Description of Internal Structures

7.3.2.1 Sand-Mud Couplets

These constitute the most characteristic point-bar successions in the mesotidal settings of the river (Smith 1988; Rahmani 1988). The sand-mud couplets are clearly visible in point-bar trench sections cut out during low tide. The sequence of laminae containing alternate sand and mud lies subhorizontal and to slightly inclined to the point-bar surface. Similar to the point-bar surface, these laminae broadly reveal a convex-up configuration. The sand-mud couplets are made up of alternate thick (2–5 cm) and thin (0.5–1 cm) sandy laminae separated by very thin mud layers (mm-scale) or mud-drapes.

Sand-mud couplets in modern tidal deposits were first reported by Visser (1980) who related its origin to the daily tidal cycles, i.e. to the flood-ebb cycles. The thicker sand layers were deposited by the dominant flood-tidal current and the thinner sandy layers by subordinate ebb-tidal currents. The mud layers or mud drapes represent sedimentation during slack water periods (Boersma and Terwindt 1981).

7.3.2.2 Large-Scale Cross-Beds with Mud Drapes and Reactivation Surfaces

Point-bar deposits locally show plannar tabular cross-bedding oriented landward, and evidently flood oriented. These cross-bedded units (30 cm thick) indicate cross-sections through megaripples and reveal a succession of tidal bundles containing reactivation surfaces, well-defined planar tabular cross-strata and mud stringers or mud laminations in response to a single tide (Boersma and Terwindt 1981). Laterally, these cross-bedded units grade to imperfectly defined cross-strata.

The reactivation surfaces represent pre-vortex structures. The foreset laminae represent full-vortex structures and the mud stringers represent suspension fall out during the post-vortex-slackening phase of a total acceleration-deceleration sequence of a single spring-tide period. The imperfectly defined cross-strata that occur laterally reflect reduced vortex activity around neap tide, when the full vortex stage did not reach the location (Fig. 7.21).

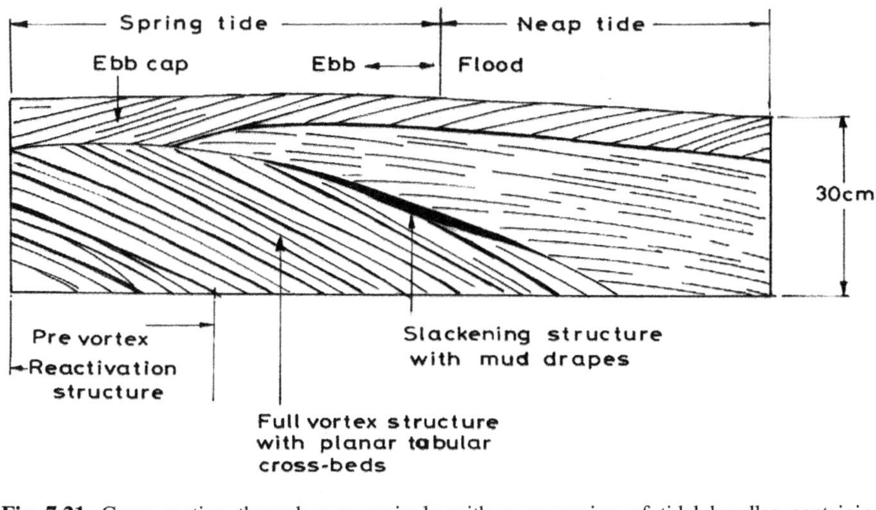

Fig. 7.21 Cross section through a megaripple with a succession of tidal bundles containing pre-vortex, full vortex and slackening structures as response of single spring tide (Field sketch)

Fig. 7.22 Kankar (concretionary) beds at Dhanchi point bar seen at a depth of 10–20 cm. The beds occur as isolated and coalescing sheets within alternating sand-mud laminae

7.3.2.3 Kankariferous (Concretionary) Laminae Alternating with Fine Sand Laminae

In some sections, a unit of sand-mud laminations is followed underneath by isolated and coalescing sheets of concretionary (Kankar) materials within sand-mud laminae (Fig. 7.22). The kankar layers are laterally discontinuous and are at most 3 cm in thickness. They occur at a depth of 10–20 cm from the surface, and their major constituents are derived from old palaeosols.

The kankar layers as channel-lag deposits are known to occur in the Chambal and Jamuna River point bars, India (Reineck and Singh 1980). Channel-lag concretionary deposits have also been described by Fisk (1944), Arnborg (1958) and Lattman (1960).

7.4 Sedimentary Structures of the Swash Platform

7.4.1 Description of Bedforms

7.4.1.1 Backwash Ripples

These are parallel-to-riverbank, gentle undulations that occur away from the river-channel margin. The bedform assumes prominence due to variations of colour from the crests to the troughs of ripples. They are sinuous in plan; wavelengths range from 36 to 62 cm, with a mean value of 47 cm and heights from 0.3 to 0.9 cm, with a mean value of 0.6 cm. The ripples are generally asymmetrical in profile with a lee-slope direction towards the river channel. Dark minerals, mostly biotite, concentrate along the troughs of ripples, whereas light–coloured, quartzo-feldspathic minerals mark the ripple crests. Thus, instead of being marked by their relief, these ripples are characterised by sinuous, alternate light- and dark-colour bands.

The backwash ripples occupy the highest topographic areas and concentrate in regions of maximum advance of wave swash within a width of 7 to 10 m. The ripples disappear as the mangrove mud banks approach. The backwash ripples migrate towards the river channel by backwash movements and reveal secondary modifications following their initial formation at higher Froude numbers (Komar 1976).

7.4.1.2 Swash Marks

These are tiny, curved ridges or markings on the sandy swash platform. The curved ridges with their convexity landward mark the maximum advance of wave swash. The ridges are generally of insignificant height (<2 mm) and show strike-wise continuation for several metres although with minor breaks in places. The ridges are generally made up of very fine sand grains.

Swash marks result from the lobate fronts of dying wave swash. In the studied area, the swash marks register the limit of the outer bank of the river channel, and, as a result their alignment is at right angles to the shoreline alignment. Hence the swash marks here occur at almost right angles to their counterparts on the sea beach.

The spacing of successive swash marks is directly dependent on the degree of slope of the beach or platform on which they occur (Emery and Gale 1951; Komar 1976). Sandy platforms with steeper slopes give rise to closer spacing of swash marks, whereas gently sloping surfaces yield wider spacing. The spacing of swash marks in the study area ranges from 5 to 20 cm and corroborates a corresponding change of slope of the swash platform from 6 to 2°.

7.4.1.3 Rill Marks

Rill marks are dendritic erosional structures on the sandy swash platform and are made by a system of small rivulets originating from a thin-layered flow of water during the falling-water stage. Rill marks are of various forms and dimensions and their morphological variations are primarily controlled by local topography, slope of the sediment surface, grain size and flowage of water.

Reineck and Singh (1980) described a number of morphological varieties of rill marks of which only three varieties are quite abundant in the study area. They are observed on slopes possessing angles of 2–6° and grain sizes ranging between that of medium sand and silt (2.4–5.6 phi).

(a) **Partially Conical Rill Marks**

They appear in the form of partially-developed conical depressions whose walls are sculptured by fine rills. The cones are around 15 cm across. Water drained from the conical rills to form larger rill marks (70–80 cm long) with accumulation tongues down slope. A sudden change in the degree of slope of the platform is marked by a change in the morphological variety of rill marks.

(b) **Bifurcating Rill Marks**

They often show downslope bifurcation and are generally sinuous or meandering in pattern (Fig. 7.23). The bifurcation is often quite strong and the last bifurcations remain open in the downslope and down-current direction. Each of the bifurcated rills moves in a meandering fashion. These rill marks are confined to a slope angle ranging from 2 to 3° and run for a distance of 3–8 m on a sandy platform.

(c) **Branching Rill Marks**

These are composed of small rill systems that bundle together to form a broad channel. The rills exhibit very prominent bifurcations to yield a dendritic pattern. The direction of bifurcation is pointed up current. The finer rills unite together downslope and are often confined to an eroded broad channel whose walls stand 3–4 cm high from the rill floor. The rill floor is also characterised by coarser lag material of fragmental shells and mud pellets. Branching rills prefer a slope angle ranging from 3 to 6°.

7.4.1.4 Rhomboid Marks

These are rhomboderal patterns or diamond-shaped structures on the swash platform. There are two different sets of rhomboid marks that appear as superimposed

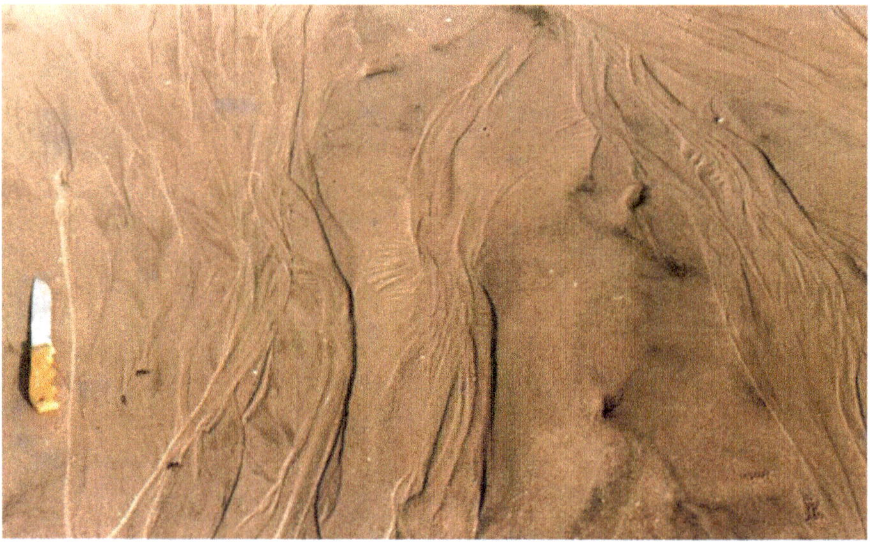

Fig. 7.23 Bifurcating rill marks. The rills show strong bifurcation in the down current direction. The flow is towards the observer. Knife is 26 cm

large-scale and small-scale reticulate patterns on the sandy surface. The smaller set ranged from 2 to 3 cm along their longer diagonals and 0.8–1.2 cm along their shorter diagonals. The larger set has longer diagonals of around 1 m and shorter diagonals around 45 cm. Rhomboid marks have positive relief of a few mm to less than 1 cm from the normal sediment surface.

Unlike their typical occurrence on sandy beach (Hoyt and Henry 1963; Komar 1976), the rhomboid marks in the study area are formed on swash platform where their long diagonals are aligned at right angles to the longitudinal profile of the river and, thus, are parallel to shoreline. Such an orientation of the long diagonals of the rhomboid marks on river-mouth swash platform apparently contradicts the longer-diagonal orientations of rhomboid marks in the adjacent beach which is at right angles to the shoreline.

7.4.1.5 Current Crescents

These are crescent shaped (U-or V-shaped) structures which widen in the flow direction. In the study area, they form around dead gastropod or pelecypod shells, and their arms open downslope towards the river direction. The crescents are 5–6 cm across around the moat and 8–10 cm along their arms. The tapering end of the dead gastropod shells always points in the down-current and downslope direction.

Current crescents in the present situation are formed from wave backwash against the dead shells strewn over the swash platform. The gastropod shells, in particular, are shifted with their tapering ends downslope due to repeated swash and backwash movement. The close association of swash marks and current crescents favours the dominance of the swash-backwash effect on the sandy platform. The opening direction of the arms of the current crescents, the concavity of swash marks and the tapering ends of the gastropod shells acting as tools all point towards the downslope direction.

7.5 Sedimentary Structures of Wash-Over Flats

7.5.1 Surface Features

A wide wash-over flat at the mouth of Thakuran reveals its channel-fill origin from the geometry of internal sedimentary structures. The intertidal wash-over flat exhibits a gradual seaward slope ranging from 3 to 6°. Further upward, the surface is characterised by intertidal ripple flat backed by a supratidal aeolian flat.

The stability of the flat is maintained to a great extent by the stilt-root system of the mangrove trees, a large number of which are stripped, leaving only portions of their upright trunks. Mangrove leaves, woods, exotic boulder and large shells of dead oysters are strewn all over the intertidal portion of the wash-over flat.

Fig. 7.24 Supratidal zone of the wash-over flat exhibiting dune-like trains due to accumulation of mangrove leaves

The oyster shells with their long axes parallel to wave swash and backwash lie embedded on the flat surface. Fine rill marks are aligned parallel to the flow directions. The supratidal zone is marked by Aeolian dune-like features produced by accumulation of dry mangrove leaves (Fig. 7.24).

7.5.2 Internal Sedimentary Structures of the Wash-Over Flat

Sedimentary units revealing a number of internal sedimentary structures have been recognised in the wash-over flat. These internal structures point to a channel-fill origin of the flat. Hummocky cross-beds generally occur at the base of the vertical sections studied and reflect a wave-induced base of the channel. Flood-generated megaripples with internal large-scale, planar, tabular cross beds follow upward, which, in turn, give way to small-scale ripples originated from reversing tidal currents and then to flaser beds. Sand-mud couplets of variable thicknesses occur at the top of most sections. A progressive decline of depositional energy is reflected in many sections. The channel-fill deposit surficially exhibits features of wash-over flats in the intertidal zone and of dune facies in the supratidal region.

Various sections cut into the wash-over flat displayed the following sequences of structural units from bottom to top:

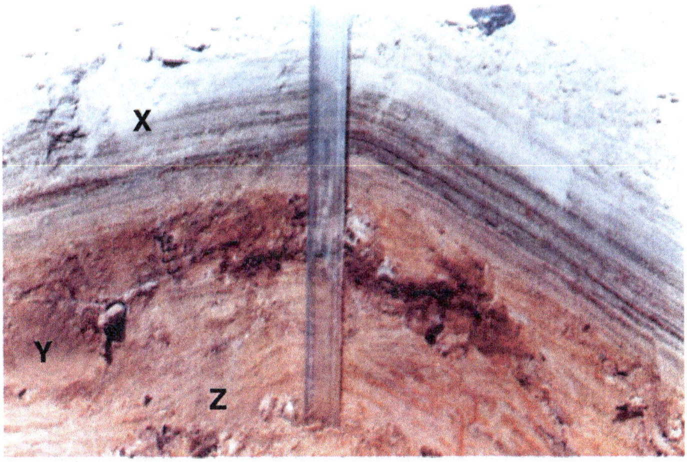

Fig. 7.25 A sequence of sand-mud couples with major breaks marked by mud laminae. The topmost part shows disruption of laminae due to sagging of sand from above (X). The middle unit is made up of drifted woody (*mangroves*) matter. The bottom of the woody zone is deformed into gentle undulations (Y). The lowermost unit reveals flaser laminations due to presence of mud lenses within sandy laminae (Z)

(i) A unit of hummocky cross-beds, followed upward by a unit of flaser beds to sand-mud couplets (Fig. 7.25).
(ii) A unit of hummocky cross-beds, followed upward by large-scale planar tabular cross-beds to a unit of sand-mud couplet (Fig. 7.26).
(iii) A unit of hummocky cross-beds, followed upward by flaser bedding into a unit of finely laminated sand.

Minor erosional plains are often present in between the units.

Fig. 7.26 Planar tabular cross-bedded unit sandwiched between overlying sand-mud alternations and underlying hummocky cross-stratifications. The foreset dips upcurrent i.e. towards the flood current. Dark zone represents woody matter. Scale is 45 cm

7.5.2.1 Hummocky Cross-Stratification

They are low angle (2°–8°), undulating cross-stratifications (Fig. 7.26). The lamina-sets (15–20 cm thick) are both concave and convex upward with wavelengths of 1–2 m and heights of 6–15 cm. The hummocks are often incomplete when traced laterally. Texturally, the hummocky cross-stratifications are made up of sand-sized particles.

These structures are generally described as being formed by storm waves (Harms et al. 1963; Davis 1983) at a depth of few metres from the wash-over flat surface.

7.5.2.2 Planar-Tabular Cross-Stratification

The planar tabular cross-beds form an isolated set in between the underlying hummocky cross-stratifications and the overlying sand-mud laminated units. The foresets dip landward at an angle of 15–18° and, thus, indicate their flood-tidal origin. The foreset laminae run for a distance of 9–10 m downdip. The cross-bedded units vary in thickness from 17–22 cm of which the individual lamina ranges in thickness from a few mm to 1.5 cm. The foresets contain mud layers punctuated by sandy laminae.

The sandy planar-tabular cross-laminated unit refers to a full vortex structure originated during flood flow. The mud drapes and mud laminae punctuated within foreset laminae refer to slackening structure. The cross-bedded unit thus represents a single spring-tide phenomenon. Occasionally wedge-shaped, partly decomposed mangrove trunks are locally trapped within the foreset laminae (Figs. 7.26 and 7.27). These are drifted woody matter from nearby mangrove forests; the wood lenses are 10–12 cm in thickness centrally.

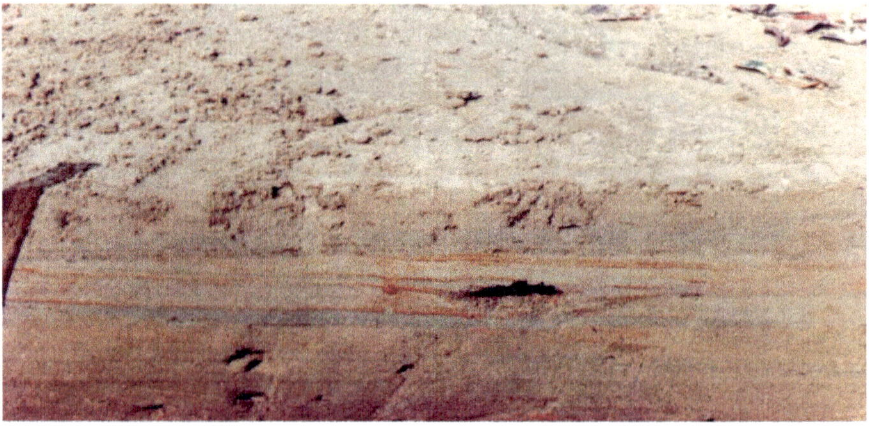

Fig. 7.27 Double sand-mud couplet in a trench section at Dhanchi. Sagging of laminae at the central part is due to bioturbation. A small tube opening is seen

The planar cross-beds are not laterally persistent. They grade to flaser beds (Fig. 7.26) and another 15–20 m away into parallel-laminated units strikewise.

7.5.2.3 Flaser Bedding

These are cross-stratified ripple bedding containing thin streaks of mud in the crests and troughs of ripples. Most flaser laminations in the study area are of wavy flaser type of Reineck and Wunderlich (1968), in which the mud flasers are concave upward when they occupy the crests of ripples and concave downward in the overlying ripple crests. In trench sections, the flaser-bedded unit is seen to continue laterally for a length of 4–5 m. The mud flasers are often of some mm in thickness (Fig. 7.25).

In general, oppositely moving flood and ebb currents with high-and-low water standstills are responsible for creating the flaser bedding (Reineck 1960). The wavy flasers, however, originate when each sand depositing current partially erodes the crests of the earlier-formed ripples and the mud streaks drape over them.

The flaser-bedded unit follows upward to an interlaminated mud-and-sand unit towards the topmost part of the sequence.

7.5.2.4 Alternate Sand-Mud Laminations

These are horizontal to subhorizontal sets of laminae with alternation of sand and mud. The set thickness varies from 6 to 15 cm in which an individual lamina is mm in thickness (Fig. 7.27). Occasionally, isolated small ripples are starved in the sandy laminae. The individual layers can be traced for 3–4 m laterally without any change in structure or granulometric properties.

The alternate sets of sand-mud layers represent deposition on more-or-less flat bottoms during the slackening of wave and current energy. Occasionally, dissipation of tractive energy leaves the ripples starved. The alternation of sets of sand-mud laminae gives a definite indication of the effects of tidal rhythms related to flood and ebb cycles.

7.5.2.5 Penecontemporaneous Deformation Structures

Trench sections on the wash-over flat locally reveal penecontemporaneous deformation structures confined within sandy units 10–15 cm thick. These units are separated by thinner muddy layers of 1–2 cm thickness. The sandy units are very commonly interlayered with mud lenses less than 1 cm thick. The deformations chiefly appear as recumbent folds with closures opposite to the dip direction of the foreset laminae (Fig. 7.28). The intensity of deformations often increases along the dip direction of the foreset laminae.

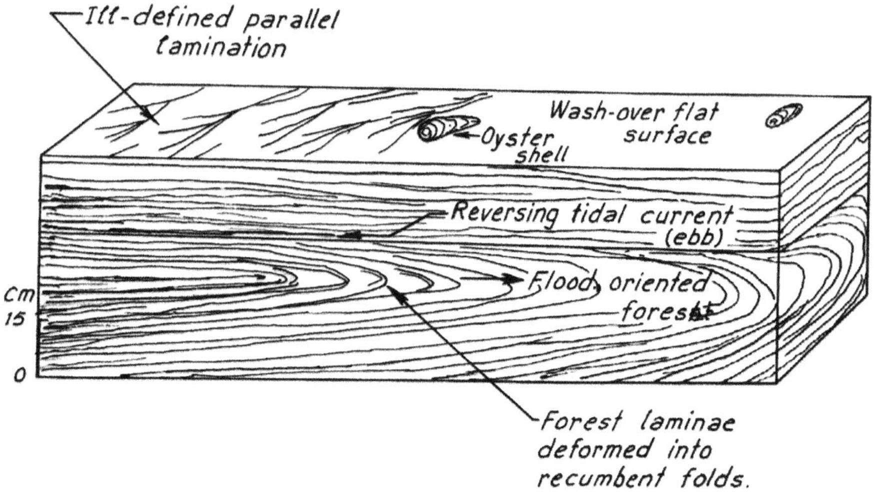

Fig. 7.28 Deformation of foreset laminae (Flood-oriented) into recumbent folds

The force of tidal currents over the water-saturated, sandy cross-laminated set exerts an overriding drag effect to produce deformation of the laminae into recumbent folds. The flowage of material simultaneous with sliding of the bounding surfaces adds to further complications in the foreset laminae along their dip directions (McKee et al. 1962a, b; Bhattacharya 1969).

7.6 Remarks on Channel Configuration and Channel-Fill

The lateral and vertical change of the internal sedimentary structures in a channel is caused, among many other reasons, by a change of bottom configuration that is deeper at the center of the channel and shallower at the margins. The large-scale planar-tabular cross-beds indicate megaripple formation along the channel thalweg. This is followed sidewise by parallel-laminated units of shallower-depth origin. The sand-mud alternations and flaser beds towards the top of the sequence point to their tidal-current origin at the dying phase of the channel. The landward dip of the foreset laminae reflects its flood-flow origin in a channel excavated primarily by flood currents.

7.7 Summary

Sedimentary structures in tidal environments have long been utilized for interpreting the hydrodynamic conditions of their formation. The effects of macrotidal regime, moderate wave energy and long shore currents have been given due importance in this regard. The various scales of tidal cycles involving the neap-spring and ebb-flood, together with wave swash and backwash, impart some modifications in the bedforms.

A quasi-equilibrium state of bedform configuration arising from unsteady flow conditions is thus a common phenomenon, and this study has revealed a full spectrum of bedform modification features in a macrotidal setting. Internal manifestation of bedforms is a direct consequences of surface features and many of these can be earmarked as features of the tidal domain. The morphology of the bedforms studied closely resembles those described in the Bay of Fundy, Canada, and Loughor estuary, UK (Table 7.2). It may be concluded that the bedforms present in the field refer to a lower flow regime of the Thakuran River; the modification bedforms are useful to interpret the total range of unsteadiness in the tidal spectrum, and areas of erosion and accretion can be well delineated from the morphological changes of the riverbeds and riverbanks. Depositional features typical of tidal environment like mud couplets, tidal bedding and tidal bundles have been recognised. Both ebb-flood and neap-spring cycles have been established from the preserved physical sedimentary structures.

Summing up, it may be concluded that the perpendicular-to-shoreline alignment of the bedforms like backwash ripples, swash marks, rhomboid marks and current crescents may be interpreted wrongly as beach features in rock records. Such a misinterpretation may lead to a 90° error in mapping a local palaeo-shoreline.

References

Allen JRL (1968) Current ripples: their relation to patterns of water and sediment motion. North Holland, Amsterdam, p 433p

Allen JRL (1984) Sedimentary structures, their character and physical bases. Development in Sedimentology. Elsevier, Amsterdam, p 663p

Arnborg L (1958) The lower part of the River Angermanalven. I Publ Georgr Inst Univ Uppsala 1:181–247

Bhattacharya A (1969) Deformed cross-beds in the point bar deposits of the River Ajay, West Bengal and Bihar. Quart J Geol Min Met Soc India. VXLI(3):169–171

Boersma JR, Terwindt JHJ (1981) Neap-spring tide sequences of intertidal shoal deposits in a mesotidal estuary. Sedimentology 28:151–170

Bogardi JL (1961) Some aspects of the application of the theory of sediment transportation to engineering problems. J Geophys Res 66:3337–3346

Boothroyd JC (1978) Mesotidal inlets and estuaries. In: Davies RA Jr (ed) Coastal sedimentary environments. pp 287–360, Springer-Verlag, New York, 420p

Boothroyd JC, Hubbard DK (1975) Coastal Engr Res Center. Misc Pap. 1–7. Army Corps of Engrs. Fort Belvoir. Virginia (1972). In: Cronin LE (ed) Estuarine Res, pp 217–234

Dalrymple RW, Knight RJ, Middleton GV (1975) Intertidal sandbars in Cobequid Bay (Bay of Fundy). In: Cronin LE (ed). Estuarine research, vol 2. Academic Press, New York, pp 293–307

Dalrymple RW, Knight RJ, Lambiase JJ (1978) Bedforms and their hydraulic stability relationship in a tidal environment, Bay of Fundy, Canada. Nature 275:100–104

Dalrymple RW, Knight RJ, Zaitlin BA, Middleton GV (1990) Dynamics and facies model of a macrotidal sand-bar complex, Cobequid Bay-Salmon River Estuary (Bay of Fundy). Sedimentology 37:577–612
Das GK (2015) Estuarine morphodynamics of the Sunderbans. Coastal research library, vol 11. Springer, Switzerland, 211p
Das GK (2016) Sedimentary structures. pp 568–572. In: Michael JK (ed) Encyclopedia of estuaries. Springer, New York, 760p
Davis Jr RA (1983) Depositional system: a geomorphic appearance to sedimentary geology. Prentice Hall, Englewood Cliffs, 669p
de Klein GV (1970) Depositional and dispersal dynamics of intertidal sandbars. J Sediment Petrol 40:1095–1127
Elliot T, Gardiner AR (1981) Ripple, megaripple and sandwave bedforms in the macrotidal Loughor Estuary, South Wales, UK. Spl Publ Int Assoc Sed 5:51–64
Emery KO, Gale JF (1951) Swash and swash mark. Trans Am Geophys Union 32:31–36
Fisk HN (1944) Geological investigation of the alluvial valley of the lower Mississippi River. Mississippi River Commission, Vicksburg, Miss, p 78p
Harms JC, Fahenstock RK (1965) Stratifications, bedforms and flow phenomena (with example of Rio Grande). In: Middleton GV (ed) Primary sedimentation structures and their hydrodynamic interpretations, vol 12. Soc Econ Paleont Mineral Spl Publ, pp 84–115
Harms JC, McKnezee DB, McCubbin DG (1963) Stratification in modern sand of the Red River, Louisiana. J Geol 71:566–580
Houthuys R, Gullentops F (1988) Tidal transverse bar building up to a longitudinal sand body (Middle Eocene), Belgium. 139–152. In: deBoer PL, VanGelder A, Nio SD (eds) Tide-influenced sedimentary environments and facies. D Reidel Publishing Company, Boston, 530p
Hoyt JH, Henry VJ (1963) The role of waves and tidal currents in the development of tidal inlet sedimentary structures and sand body geometry: examples from North Carolina, South Carolina and Georgia. J Sed Petrol 49:1073–1092
Knight RJ (1972) Cobequid Bay sedimentology project: a progress report. Mar Sedi 9:45–60
Komar PD (1976) Beach processes and sedimentation. Prentice Hall Inc., New Jersey, p 429p
Lattman LH (1960) Cross-section of a flood plain in a moist region of moderate relief. J Sedi Petrol 30:275–282
McKee ED, Reynolds MA, Baker CH Jr (1962a) Laboratory studies on deformation on unconsolidated sediments. US Geol Survey Profess Papers 450-D:151–155
McKee ED, Reynolds MA, Baker CH Jr (1962b) Experiments on intraformational recumbent folds in cross-bedded sand. US Geol Survey Profess Papers 450-D:156–160
Potter PE, Pettijohn PE (1963) Palaeocurrent and basin analysis, 2nd edn. Springer-Verlag, New York, p 423p
Rahmani RA (1988) Estuarine tidal channel and nearshore sedimentation of a Late Cretaceous epicontinental sea, Drumheller, Alberta, Canada. pp 433–471. In: deBoer PL, VanGelder A, Nio SD (eds) Tide-influenced sedimentary environments and facies. D Reidel Publishing Company, Boston, 530p
Reineck HE (1960) Uber Zeitlucken in rezenten Flachsec—Sediment Geol Pundschan 49:149–161
Reineck HE (1972) Tidal flats. In: Rigby JK, Hamblin WK (eds) Recognition of ancient sedimentary environments, vol 16. Soc Econ Paleon Mineral Special Publication, pp 146–159
Reineck HE, Singh IB (1972) Genesis of laminated sand and graded rhythmites in storm-sand layers of shelf mud. Sedimentology 18:123–128
Reineck HE, Singh IB (1980) Depositional sedimentary environments. Springer Verlag, New York, p 408p
Reineck HE, Wunderlich F (1968) Classification and origin of flaser and lenticular bedding. Sedimentology 11:99–104

Sanders JE (1963) Concept of fluid mechanics provided by the primary sedimentary structures. J Sediment Petrol 31:446–462

Smith DG (1988) Modern point bar deposits analogous to the Athabasca Oil Sands, Alberta, Canada. pp 417–432. In: deBoer PL, VanGelder A, Nio SD (eds) Tide-influenced sedimentary environments and facies. D Reidal Publishing Company, Boston, 530p

Terwindt JHJ (1988) Paleo-tidal reconstruction of inshore tidal depositional environments. pp 233–264. In: deBoer PL, VanGelder A, Nio SD (eds) Tide-influenced sedimentary environments and facies. D Reidel Publishing Company, Boston, 530p

Van Straaten LMJU (1961) Sedimentation in tidal flats areas. J Alberta Soc Petrol Geol 9:203–213

Visser MJ (1980) Neap-spring cycles reflected in Holocene subtidal large scale bedform deposits: a preliminary note. Geology 8:543–546

Chapter 8
Bioturbation Structures

Abstract Bioturbation structures produced by the interaction of living organisms and soft sediments manifest themselves in a wide range of forms like tracks and trails, footprints, burrows, mounds and pellets in various geomorphic domains of the meso-macrotidal ThakuranRiver of Sunderbans. Study of behavioural pattern of these organisms through bioturbation structures helps geologists in deciphering the environmental conditions prevalent during the intervening period of deposition of soft sediments and their lithification. They may be of lesser magnitude, but the processes of bioturbation have the potential to alter depositional sequences and sediment chemistry through the process of exhumation and subsurface mixing of sediments, as evidenced in cases of burrows in the intertidal mudflats of the Thakuran River.

Keywords Bioturbation structures · Track · Trails · Foot prints · Fin marks · Mounds · Press marks · Burrows · Tubes · Pits · Borings · Thakuran river · Sunderbans

Organization of sediments by physical processes in the tidal environment and the physical sedimentary structures formed thereby have been observed as common phenomena from the sedimentological point of view in the coastal environment. But disturbance-induced modifications in the sedimentary deposits due to the activity of living organisms seem to be no less important in the exposed and sheltered intertidal zones of the deltaic Sunderbans. Intertidal zones of the Thakuran River including river banks, natural levees, point bars and mid-channel bars inundated twice daily in the semi-diurnal situations (Das 2015) are scattered among several bioturbation structures.

As defined by Frey (1975), bioturbation structures or libensspurensare biogenic sedimentary structures that disrupt physical stratifications or rearrange the sediment fabric by the activity of organisms in the form of tracks, trails, burrows and similar structures. These features are described as 'trace fossils' when recorded in ancient rocks. As these structures are produced by the activity of living organisms within or on the sediments, they sculpt various patterns on the sediment surface and may even

G.K. Das, *Tidal Sedimentation of the Sunderban's Thakuran Basin*,
DOI 10.1007/978-3-319-44191-7_8

lead to complete or partial destruction and reformation of physically-produced primary or secondary structures.

The density of animal population vis-a-vis density of bioturbation structure depends mostly on: (i) oxygen content of the substrate and of the ambient environment based on water circulation, (ii) availability of food, (iii) grain size controlled by hydrodynamic conditions and (iv) conditions of sedimentation (Hertweck 1975). The role of animal-sediment interactions, biogenic alteration of recent marine and freshwater deposits as well as of their ancient counterparts have been excellently summarised by McCall and Tevesz (1982). Frey and Howard (1969), Hertweck (1975) and Basan and Frey (1977) differentiated various coastal zones based on bioturbation structures. The classical work of Frey (1975) on trace fossils provides an important document for working with bioturbation structures of both recent and recent origin.

8.1 Basis of Classification of Bioturbation Structures

Various attempts have been made to classify bioturbation structures as per the need and purpose for which they have been employed (e.g. Reineck and Singh 1980). In the present work, the author prefers to classify the various bioturbation structures into a three-fold classification after Hertweck (1970). These are: (a) surface libensspuren, (b) internal libensspuren and (c) dwelling structures. All the bioturbation structures encountered in the various sub-environments of the study area can be conveniently grouped as per the above scheme (Figs. 8.1 and 8.2). The author is, however, aware of the critique of Reineck and Singh (1980) regarding the demerits of this classification to differentiate the structures. Reineck and Singh (1980) argue that, as the benthic animals may move within sediments along the bedding planes, the internal bioturbation structures may be misinterpreted as surface-bioturbation structures in ancient sedimentary rocks if the bedding planes becomes exposed after erosion. This limitation, however, is not applicable to the present study as it is employed in the recent sedimentary deposits where surficial and internal processes can be easily distinguished.

The large variety of bioturbation structures, when analysed from the sedimentological point of view, reveals that the shape of organisms, the activity of their functional organs and the behaviour of organisms determine the actual details of the bioturbation structures. Furthermore, the animals have their preferential habitats based on sandy and muddy substrates. Accordingly, the nature of biogenic bed roughness, viz.mounds, tubes, trails, pits, etc., is the surficial expression of the nature of bioturbational activities in sediments. Table 8.1 depicts the interrelations among sedimentary facies, bed roughness and the bioturbation structure of the study area of the Thakuran River of the Sunderbans.

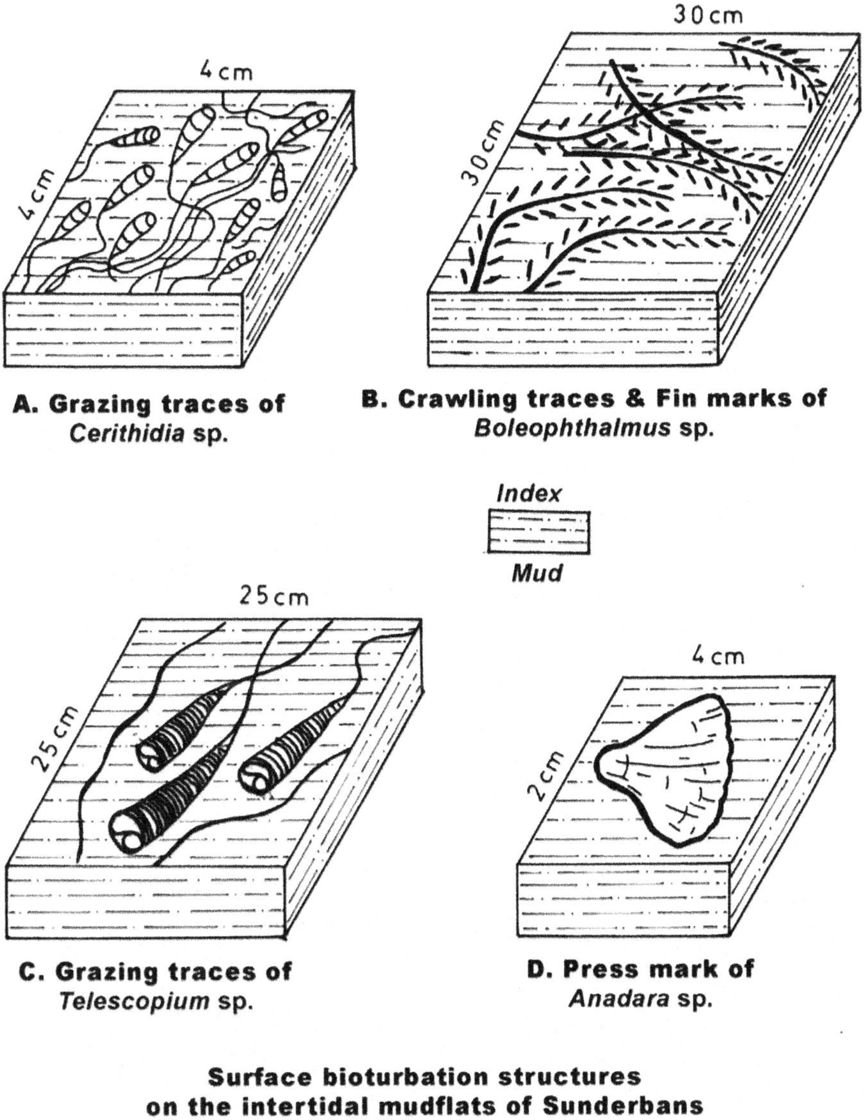

Fig. 8.1 Surface bioturbation structures on the intertidal mudflats of Thakuran River

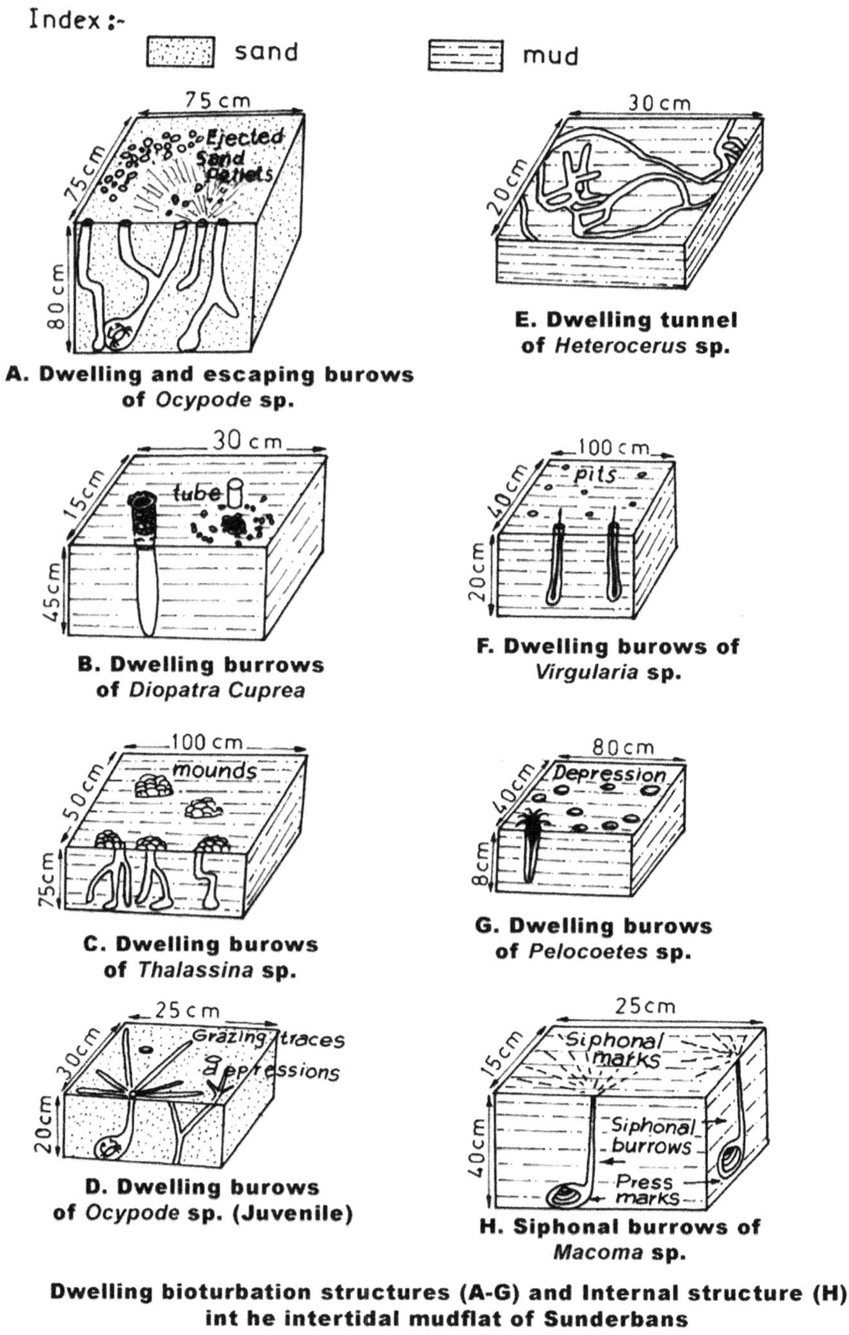

Fig. 8.2 Dwelling bioturbation structures in the intertidal mudflat of Thakuran River

Table 8.1 Sedimentary facies and bed roughness and their relationships with bioturbation structures

Sedimentary Facies	
I. Mud Facies (Silt-92 %, Clay-8 %) A. Increasing bed roughness = considerable topographic relief 1. Mounds (around or adjacent burrow openings) (i) *Uca* sp around feeding burrow (ii) *Ocypode* sp around dwelling burrow (iii) *Macrophthalmus* sp around dwelling burrow (vi) *Thalassina* sp around dwelling burrow (v) *Metaplax* sp around dwelling and escaping burrow 2. Tubes (raised up to 1 cm or more from surface) (i) Feeding tube of *Diopatra* sp (ii) Resting tube of *Ophiactis* sp 3. Trails Ridges and grooves by (i) Grazing of *Ocypode* sp (ii) Crawling of *Telescopium* sp (iii) Crawling of *Cerithidea* sp (iv) Browsing of *Carcinoscorpius* sp (v) Fin marks of *Boleophthalmus* sp 4. Pits/depression/press mark (i) Pits of *Macoma* sp (ii) Depression of *Paracondylactis* sp (iii) Depression of *Pelocoetes* sp (iv) Press mark of *Anadara* sp (v) Pits of *Cloridopsis* sp (vi) Pits of *Virgularia* sp 5. Tracks (i) Foot prints of birds (ii) Hoof marks of *Axes axes* (Deer) (iii) Foot prints of *Panthera tigris* (Royal Bengal Tiger) 6. Tunnels (i) Subhorizontal tunnel by *Carcinoscorpius* sp (ii) Horizontal tunnel of *Heterocerus* sp B. Decreasing bed-roughness = smooth surface topography (homogenize material by reworking) formed by small and mobile animal (i) Pits and depression of *Ocypode* (juvenile) sp (ii) Pits of *Donax* sp	II. Sand Facies (Coarse sand-12 %, medium sand-82 % and fine sand-6 %) 1. Burrows of *Ocypode* sp with scattered ejected pellets 2. Pits of *Uca* sp 3. Depression with marginal rim of *Paracondylactis* sp 4. Grazing marks and pellets of *Dotilla* sp

8.2 The Classification

8.2.1 Surface Bioturbation Structures

Bioturbation structures that are recognisable on the surface of sedimentary deposits can be sub-divided into the following types based chiefly on ethology, i.e. on the behaviour of organisms as deduced from their traces.

8.2.1.1 Tracks: Discontinuous Marks of Footprints Left on Sandy and Muddy Surfaces (e.g. Bird Foot Prints, Deer Hoof Marks and the Foot Prints of Tigers)

(a) **BirdFoot Prints**

The Thakuran basin of the Sunderbans is famous for the stop-over of certain avian communities. These birds often move on the sandy and muddy surfaces of the intertidal deposits in search of food. As a result, birdfoot prints are quite common in areas where there is greater population of benthic-animal communities. The foot prints of birds may continue forfew metres to several hundred metres and may be rectilinear to curvilinear in pattern. The tracks maintain a steady trend without showing much dislocation.

(b) **HoofPrints of Deer (*Axes axes*)**

The intertidal mudflats in and around the riverbanks, with dense mangrove vegetation in the Dhanchi reserve forest, bear evidence of a number of foot print of Sunderbans deer in the form of trackways (Frey 1973). These are hoof marks where a pair of laterally displaced marks is imprinted on the muddy surface. The hoof-marks are characterised by a subelliptical depression with a central ridge along the direction of locomotion. The hoofs are 8 to 10 cm long and 5 to 6 cm broad. The stride is 55–60 cm and indicates the direction of movement of the track maker.

(c) **Foot Prints of Royal Bengal Tigers (*Panthera tigris*)**

The upper mudflat of the Bulchery Island revealed a rare exhibit of the foot prints of royal Bengal tigers association with those of some other quadrupeds (e.g. fox and large mongoose). As only one foot print of the tiger was present on the mudflat, it was not possible to measure the stride and pace of the track maker. The imprint is 24–25 cm long and 14–16 cm wide with five distinct impressions of the paw. The rectangular nature of the large impressions and the elongated nature of the finger prints indicate that the animal was female. Sarjeant (1975) has given an elaborate description of vertebrate foot prints together with their implications in determining the nature of their movement, i.e. walking or racing manner.

8.2.1.2 Ecological Importance of Foot Prints

The Sunderbans Biosphere Reserve is the abode of several vertebrate and invertebrate animals. This complex ecosystem is also characterised by a network of food webs where the flow of energy in terms of predator-prey relationship is well documented by a number of workers (Sanyal 1994; Chakraborty 1994). The close association of the imprints of Sunderbans deer and that of the royal Bengal tiger revealed a unique predator-prey relation which is deduced from sedimentological evidence. Thus, the importance of the study of bioturbation is not only restricted to animal-sediment interactions, but also this study helps in understanding the transference of energy in the food chains of the inhabiting animals.

8.2.1.3 Crawling Traces: Trail Marks Left on Sediments Due to Locomotion of Organisms

Animals having part of their bodies in contact with the substrate surface produce continuous grooves during locomotion. The evidence may also appear as sub-surface traces made by animals moving from one place to another. The tracks are formed by walking bristles or other appendages or be caused by muscular movements of the body on both sandy and muddy substrates. It has been observed that the same animal performs various movement patterns like walking, cantering or galloping, all of which create different crawling traces (Simpson 1975).

The muddy surface of the mid-channel bars and point bars of the Hooghly-Matla estuarine complex are profusely marked by the sinuous crawling traces of *Telescopium telescopium*. These traces continue for over 300 cm in many instances. The widths of the traces corroborates with the animal width at its base which ranges from 3 to 5 cm. The locomotive trace of *Telescopium telescopium* registers a gradual tapering towards its distal end.

The mud skippers *Boleophthalmus* sp., on the contrary, make a number of straight to slightly wavy traces that radiate from a central burrow opening which is often used as a refuge. These traces are produced by the movement of the ventral part of the animal's body during locomotion. A series of biserial pectoral fin marks also ornament the central markings on two sides. The traces make a jumbled mess around the central burrow. The traces of *Cerithidea* sp. may impart further complications in the architecture. In other instances, the sinuous, elongate trails of certain unknown gastropods characterise the mudflats of the Thakuran River, particularly on the creek margins. These markings are sinuous grooves on muddy surfaces and promote draining of interstitial water.

Crawling traces and trails are also frequent on the sand flats of swash platforms and mid- channel-bar surfaces. Figure 8.3 exhibits the trace of a gastropod commonly known as moon shell (*Polinices* sp.) on a sandy-swash platform of the Thakuran River. The animal is seen to rest at the end of the trace, which is formed

Fig. 8.3 Crawling trace of the moonshell *Polinices* sp. (Gastropod) on the sandflat of swash platform. The animal is seen at the end of the trace. The marginal ridges are quite prominent. The pen is 15 cm

during locomotion of the animal. The central groove with elevated ridges on either side leaves a sandy trail on the surface.

8.2.1.4 Browsing Traces: Marks Left by Organisms Feeding on Sediment Surfaces

These traces are formed by excavations made by deposit feeders while in search of food within, or on, the sediment surface. *Metaplax* sp. forms a radial pattern during browsing; these radially-arranged browsing traces vary in number from two to six in many instances. The traces diverge from a central burrow and are straight to slightly wavy in nature. On the other hand, browsing traces formed by *Carcinoscorpius rotundicauda* (King crab) are loop-like to very irregular. Very fine ridges and grooves run within the trace boundaries. The width of the browsing traces is dependent on the width of the animal responsible for creating the markings. As the grooves are deep enough on most occasions, they have high potential to be preserved and can be seen from ancient rock records in the form of sole marks at the bottom of overlying sedimentary units. Browsing scratches on sediment surfaces further augment draining of interstitial water in the form of fine rills at late-stage emergence.

8.2.1.5 Grazing Traces: Marks Left on Sediments Depending on Grazing Behaviour

Grazing traces are often made by gastropods. The traces are shallow to deep on sediment surfaces, depending on the size and weight of the grazers. The traces are sinuous and vary in length from one centimetre to tens of centimetres. Generally a 'herd' of *Cerithidea cingulata* grazing on the algal mats of mudflats imparts surface roughness. An assemblage of several grazing marks imposes a complex pattern on the sediment surface. Occasionally, 'nibbing' by *Cerithidea* sp. terminates abruptly at the sharp edge of algal mats. Crawling traces of *Telescopium telescopium* and resting refuges of *Anadara* sp. introduce yet more bed roughness (Fig. 8.4). These animals are closely associated with *Cerithidea* sp. move parallel to create parallel markings. The animals rest at the end of each groove, and this helps identifying the direction of movement of the animals.

Among the grazers, the amazingly fast-moving ghost crabs *Ocypode macrocera* constitute a very important community. They produce surface roughness by making grazing scratches and heaping pellets on sand flats.

It is worth mentioning in this connection that although algal mats help restore stabilization of sediment surfaces, biogenic scratching, nibbing and grooving destabilize the textural bondage. This is well exhibited by the various grazing marks produced by the deposit feeders on the intertidal sand and mudflats.

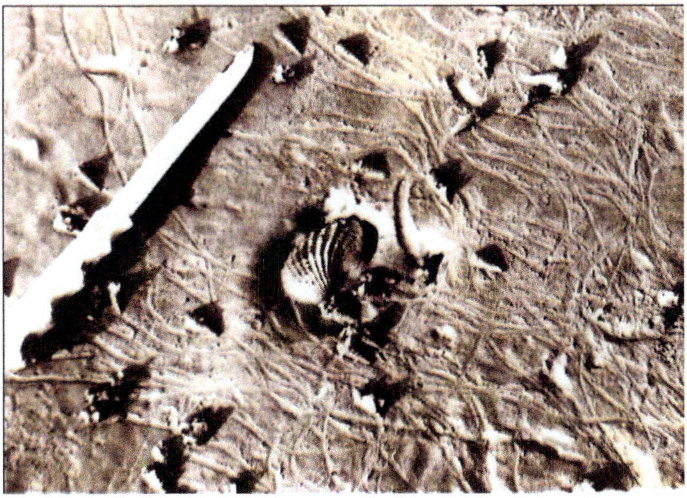

Fig. 8.4 An *Anadara granosa* is at rest on the mudflat. Nibbing of *Cerithidia* sp. makes a mess all around. Note the random orientation of *Cerithidea cingulata*. Knife is 26 cm

8.2.1.6 Resting Traces: Marks Left by Organisms Resting on Sediment Surfaces

These impressions are produced by animals while they are at rest on sediment surfaces. Resting traces of hard molluscan shells are common and exhibitan impression of the valve of *Anadara granosa* on the mid-channel-bar mudflat surfaces of the Thakuran River. The valve impression is 3 cm long and 2.6 cm wide. The quadrangular shape of the valve with radial growth lines is clearly seen. The left portion of the impression is superimposed by the fin marks of *Boleophthalmus* sp.

In some instances, scours are formed around the resting place of the animals due to the circulatory motion of tidal water. Eventually the animals get incised on the sediments or may lie in some depression. The close association of *Cerithidea* sp. and *Anadara* sp. or *Boleophthalmus* sp. and *Anadara* sp., together with their functional markings, help interpreting the paleoecology when the animals and their activities are preserved properly.

The other resting trace of importance on the muddy point bars of tidal creeks is of *Carcinoscorpius rotundicauda* (king crab). The animals are quite large. Their length ranges from 15 to 20 cm and breadth from 12 to 15 cm. The telson length is about 20 cm. The impression of the margins of the animal, together with the whole or part of the telson impression in the form of a linear depression, is often well preserved on the muddy surfaces (Fig. 8.5).

Fig. 8.5 A *Carcinoscorpius rotundicauda* (King crab) is at rest on the mud flat. The telson marking is quite prominent. Scale is 15 cm

The resting traces may be converted to dwelling structures depending on the change in the modes of activity of the animals (Osgood 1970). Figure 8.5 illustratesthe burrowing activity of *Carcinoscorpius* sp., where the sediments are dug out to form surficial mounds.

8.2.1.7 Fin Marks: Traces Produced by Swimmers like Fish While Moving on Sediment Surfaces

Intertidal flats of the Thakuran River are very commonly dominated by mud skippers like *Boleophthalmus* sp. (Gobid fish). Pectoral fin marks produced by *Bolephthalmus* sp. are quite common. The well-preserved nature of fin marks on muddy surfaces is thought to be due to the cohesive nature of mud reinforced by microbial matting. In a different area, on the lower part of a point bar near Damkal, the muddy surface has inscribed by numerous shallow and short criss-cross fin marks of some unidentified swimming animals.

8.2.1.8 Mounds: Heaps of Granular or Pelletoidal Materials Accumulated at Burrow Mouths

These vary in shape and size. The mounds may be conical domes, regular crater-like bodies to irregular in shape. These are generally produced by the ejection of granular or pelletoidal materials at the time of burrowing by ghost shrimps (*Thalassina* sp.; Fig. 8.6) and fiddler crabs (*Uca* sp.). The mounds are commonly observed on mud flats of the upper intertidal zone. The mound materials are mostly muddy (M_Z = 6.12–6.40 phi). Figure 8.6 exhibits a small domical mound (dia: 13 cm; ht: 6 cm) formed by a ghost shrimp at a Dhanchi mid-channel bar. The mound slopes uniformly on all sides from its apical region having a central exhalation pit. Close to this mound, a separate pit at the centre of a depression is also seen clearly. This pit is perhaps used for inhalation purposes.

Fig. 8.6 A small domical ghost shrimp mound with a central pit at the apex. Note a separate pit at the centre of a nearby depression. Algal matting is clearly visible on the mudflat. Knife is 26 cm

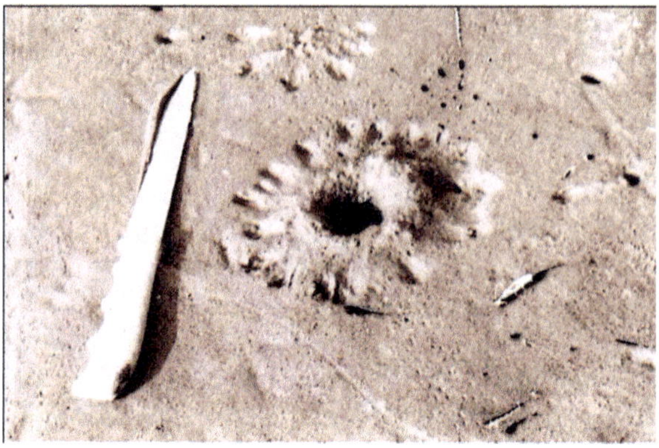

Fig. 8.7 Crater-like mound around exhalent burrow opening of *Metaplax crenulata* on the mudflat. Note the flowering pattern around the burrow. The peripheral discontinuity is due to the movement of the animal from the burrow opening to all directions

Some mounds of the fiddler crab, *Metaplax crenulata*, form a typical crater-like structure with a central opening. On many occasions, these cratered mounds are discontinuous at their peripheral margin, and this is due to removal of mound material along the paths of movement of the animal from the burrow mouth in all directions (Fig. 8.7).

The burrowing crustaceans thus act as agents for recycling sediments both laterally and vertically. The net sediment budget remains same, and the intertidal materials show thorough mixing in and around the burrows. Biogenic subduction (Aller 1982) is always associated with the process of burrowing.

8.2.2 Internal Bioturbation Structure

Internal bioturbations are impressions produced within the sediments and are formed mainly by animals living in, and/or free moving within, sediments. They are more prominent in tidal mud flats of both accretionary point bars and mid-channel bars.

8.2.2.1 Press Structures: Impressions Produced by Animals Having Restricted Movement in and Around Their Dwelling Places Within the Sediments

Press structures are rather rarely observed as the animals producing such structures do not leave many surface indications. However, this structure is well exhibited by *Macoma* sp. in the mudflats of point bars and mid-channel bars. *Macoma* sp. makes more or less vertical siphonal burrows of a few millimetres in diameter and produces press marks at a depth of 40–45 cm where they rest. Their presence at depth is evidenced by a flowery structure around their siphonal burrow, formed during grazing by the siphons.

8.2.3 *Dwelling Structures: Burrows and Tubes Made by Organisms in Which They Live or Use for Escape*

Bioturbational activities during digging are performed for the purpose of dwelling and escaping and feeding. The dwelling and escaping burrows are generally simple, straight or 'U' shaped—whereas, feeding burrows are much more complicated. The dwelling structures can be of three types as explained next.

8.2.3.1 Borings: Burrows in Hard Substrates

These are observed mostly on dead mangrove trunks or on drifted plant materials that rest on the sediments. The inner parts of the borings made by polychates are generally of 2–6 mm diameter, and often contain hard calcified linings in the inner walls. These borings, in turn, act as refuges for fiddler crabs, brachuran crabs, sea anemones, tube anemones and sea pens.

8.2.3.2 Dwelling Burrows on Sand and Mud Deposits: Used for Both Dwelling and Escape Purposes

The burrows may be short and simple (e.g. of sea pens and sea anemones), straight and tubular with coatings of parchment-like materials (e.g. of Polychaetes), unbranched, branched, Y-shaped or U-shaped (e.g. of Ghost crab, *Ocypode* sp., and fiddler crabs, *Uca* sp.). The burrows are of variable length (3.2–8.4 cm) and width (0.05–3.5 cm). The insect like *Heterocerus* sp. (Coleoptera) creates very narrow, sinuous to intricately complex 'tunnels' on intertidal muddy surfaces (Fig. 8.8). A detailed list of the macrobenthic burrowing animals and the nature of their burrows are given in Table 8.2.

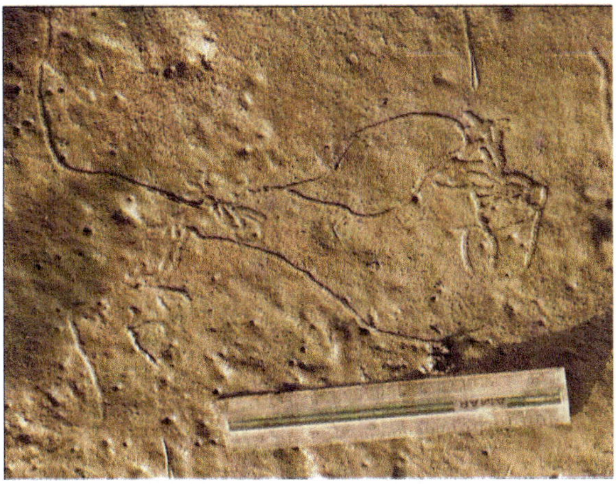

Fig. 8.8 Intricate and narrow tunnels of *Heterocerus* sp. on the mudflat. Scale is 15 cm

Table 8.2 List of macrobenthic burrowing animals and the burrow characters in the deposits of Hooghly–Matla estuarine complex

Burrowing organism	Geomorphic Location and bed materials	Burrow Character				Associated Surficial Mounds/rims/tubes/depressions/pits/pellets
		Length (cm)	Diameter (cm)	Dip and interconnection	Type	
Ph: CNIDARIA A. SEA ANEMONE *Paracondylactis* sp.	Intertidal mud flats and sand flats of point bars and mid channel bars	20–28	1.0–3.5	Vertical, non-connected	Dwelling	Depression with marginal rim
Pelocoetes exul	Intertidal mud flats of point bar and marginal bar	8–10	0.8–1.4	Vertical, non-connected	Dwelling	Circular depression
B. SEA PEN *Virgularia* sp.	Intertidal mudflats of marginal bars	15–18	0.05–0.1	Ventral, non-connected	Dwelling	Pits
Ph. ANNELIDA A. POLYCHAET *Diopatra* sp.	Intertidal mud flat of point bars, marginal bars, mid channel bars	25–53	1.0–1.3	Inclined zig-zag, non-connected	Dwelling and feeding	Tube
Ph. ARTHOPODA A. CRABS *Metaplax crenulata*	Intertidal mudflats of point bars, mid channel bars, marginal bars	20–30	1.8–3.5	Vertical, non-connected	Dwelling	Marginal rims; sometimes with peripheral discontinuity
Ocypode macrocera	Supratidal to upper inlet sand flats of mid channel bars, point bars	35–80	1.5–6.0	Curved, L-shaped and inter connected	Feeding and Dwelling	Scattered pellets in and around the burrow mostly along feeding spurs covering 95°–180°
Uca acuta	Supratidal muddy, natural levee	15–20	1.2–3.5	Curved, slightly inclined	Dwelling	Mounds around burrow
Macrophthalmus sp.	Supratidal mudflats of point bars, marginal bars and river banks	28–35	2.1–3.3	Inclined, curved and L-shaped; inter connected	Feeding	Marginal rims; nearly circular pellets form irregular mounds
Dotilla sp.	Silty sand flats; point bars and marginal bars	3.2–9.1	0.3–0.6	Vertical, non-connected	Dwelling and Feeding	Pits and pellets

(continued)

Table 8.2 (continued)

Burrowing organism	Geomorphic Location and bed materials	Burrow Character				Associated Surficial Mounds/rims/tubes/depressions/pits/pellets
		Length (cm)	Diameter (cm)	Dip and interconnection	Type	
B. GHOST SHRIMP *Thalassina anomala*	Intertidal marshy mud flats of point bars, mid channel bars	55–90	0.4–0.7	Vertical, non-connected	Dwelling	Conical mound, sometimes with auxiliary cones
C. SQUILLA *Cloridopsis bengalensis*	Intertidal mud flats of point bars	14–23	0.1–0.4	Horizontal, straight; inter connected	Dwelling	Pits
D. INSECT *Heterocerus* sp.	Intertidal mud flats of point bars, marginal bars	60–85	0.06–0.08	Horizontal to sub-horizontal, inter connected; complex network	Dwelling	Horizontal tunnels
Ph. MOLLUSCA A. BIVALVE *Macoma birmanica*	Intertidal mudflats of point bars, marginal bars, mid channel bars	25–51	0.6–0.8	Vertical, slightly inclined, non-connected	Escape, feeding, siphonal	Depression around burrow, surface with flowery star shape marks by foraging siphon
Donax sp.	Intertidal mudflats	4–6	0.2–0.5	Vertical	Siphonal	Small pits
Ph. ECHINODERMATA A. OPHIUROIDS *Ophiactis* sp.	Intertidal mudflats of mid channel bars and marginal bars	5–10	0.05–0.08	Vertical, non-connected	Dwelling and resting	Small pits

The sand flats of mid-channel bars and swash platform near the river mouth register profuse bioturbational activities. The animals disturb the deposits both internally as well as surficially. The dwelling burrows of ghost crabs (*Ocypode macrocera*) heavily disturb the internal laminations and create a number of surface structures both by ejecting ingested sandy materials and pelletization.

Digging operations by *Ocypode macrocera* in the upper-intertidal sand flat of the Dhanchi mid-channel bar are often seen during low tides. The animals strike the sandy substratum repeatedly to impose a quick-sand-like property to the sediments for the purpose of digging. These areas are commonly strewn by numerous spherical sand pellets 8 to 10 mm in diameter.

In many instances, radially extended feeding spurs are formed on the surface during the feeding of *Ocypode* sp. The spurs cover 95–180° of arc of a circle. The feeding lines are straight and extend up to 50 cm (Fig. 8.9). Clots of ingested materials ejected by the crabs in their process of feeding and digging are relatively large and without any well-defined shape. These clots mark the outer limits of the feeding zone. In many instances intensive burrowing by crab gives a very-complex mottled look to the bank sediments.

The fiddler crab, *Uca* sp., acts as a very important agent of bioturbation. They make tubes of granular aggregation of mud, which are often exposed for 2 cm above the ground level. The thicknesses of the tube walls vary from 3 to 5 mm while the diameters range from 0.8 to 1.0 cm. The tubes often collapse by tidal action, and the tube-building granules remain scattered around the centres of the tubes.

Sandy intertidal zone sediments are often manipulated by *Ophiactis* sp. (Echinoderm), which may penetrate some 10 to 12 cm below the surface. They produce very narrow (<3 mm in diameter), straight, vertical burrows. The burrows are parallel to each other and are essentially restricted to the muddy zone.

The role of polychaetes such as *Nereis* sp., *Perinereis*sp and *Diopatra cuprea* create semi-permanent dwelling tubes of variable length (15–30 cm) and diameter

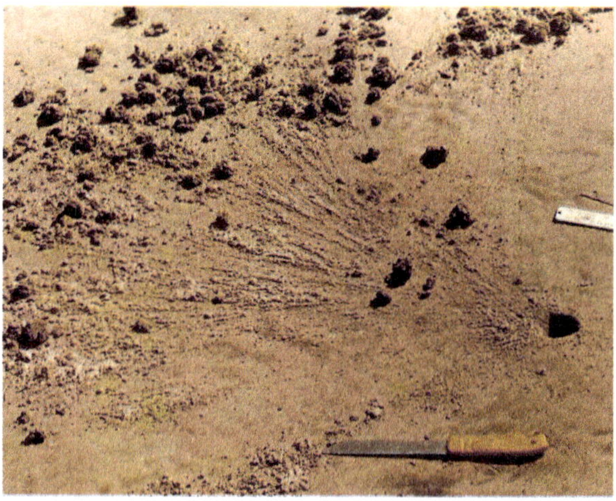

Fig. 8.9 Radially extended feeding spurs of *Ocypode macrocera* on the sand flat. Burrow opening and ejected pellets are clearly seen. Knife is 26 cm

(3 mm–5 cm) in the intertidal mudflats. Generally these semi-permanent tubular bodies are made up of fine sand and mud with very small pieces of shells and organic debris cemented together by mucus materials. In many instances, the exposed portions of the tubes remain intact above the surface and stand as obstacles in the path of moving water.

8.2.3.3 Feeding Burrows: Interconnective Burrows Made by Animals for Feeding

These are often complicated in pattern. Sometimes the deposits are mottled by numerous burrows of 5–40 mm in diameter. The lengths of burroware often difficult to measure because of their frequent interconnections. At Dhanchi, the supratidal mud banks exhibit clusters of ejected food pellets of *Macrophthalmus* sp.

8.3 Summary

Dwelling, feeding and internal bioturbation structures have profound roles in churning the mudflat and sand-flat sediments. Very commonly, an overall mixing of sediments takes place, and this alters the textural properties of sediments deposited by physical processes. The primary depositional behaviour of sediments produced by swash-backwash (waves) effects and by tidal currents is manipulated largely by different intensities of biogenic activity. Since the net sediment budget of sand and mudflats remains constant, burrowing leads to both lateral and vertical exchange of sediments and biogenic subduction of surface topography. The internal laminations are disturbed to different degrees and are deformed by folding and microfaulting. The ejected burrow materials in the form of mounds, castings and projected tubes commonly create complications in the surface-flow patterns and flow directions. Flow divergence against these mounds and obstacles becomes responsible for the generation of current crescents and the modification of bedforms.

References

Aller RC (1982) The effects of macrobenthos on chemical properties of marine sediments and overlying water. In: McCall PZ, Tevesz MJS (eds) Animal sediment relations. Plenum Press, New York, pp 53–102

Basan PB, Frey RW (1977) Actual palaeontology and neoichnology of salt marshes near Sapelo island, Georgia. In: Crimes TP, Harper JC (eds.) Trace fossils 2 (Geol. J Spec Issue 9), Seel House Press, Liverpool, England, pp 41–70

Chakraborty K (1994) Biodiversity of mangrove ecosystem of Sunderbans. In: Maity S, Mukherjee N (eds) Bidhan Chandra Krishi Viswavidyalaya Publ, pp 57–64

Das GK (2015) Estuarine morphodynamics of the Sunderbans. Coastal research library, vol. 11. Springer, Switzerland, 211p

Frey RW (1973) Concepts in the study of biogenic sedimentary structures. J Sediment Petrol 43:6–19

Frey RW (1975) The realm of ichnology, its strength and limitations. In: Frey RW (ed) The study of trace fossils. Springer, Switzerland, p 13–38 (562p)

Frey RW, Howard, JD (1969) A profile of biogenic sedimentary structures in a Holocene barrier island–salt marsh complex, Georgia, Gulf Coast Assoc Geol Socs Trans 19:427–444

Hertweck G (1970) The animals community of a muddy environment and the development biofacies as effected by the life cycle of the characteristic facies. In: Crimes TP, Harper JC (eds) Trace fossils (Geol J Spec Issue 3), pp 235–242

Hertweck G (1975) Ichnology aspects. In: Dorjes J, Hertweck G (eds) Recent biocoenoses and ichnocoenoses in shallow water marine environments. In: Frey RW (ed) The study of trace fossils. Springer, Heidelberg, pp 468–491

McCall PL, Tevesz MJS (eds) (1982) Animal sediment relations. Plenum Press, London, p 336p

Osgood RG Jr (1970) Trace fossils of the Gincinnat area. Palaeontographica Amer 6(41):281–444

Reineck HE, Singh IB (1980) Depositional sedimentary environments. Springer, New York, 408p

Sanyal P (1994) Mangrove distribution in Indian Sunderbans. In: Maiti S, Mukherjee N (eds) Bidhan Chandra krishi Viswavidyalaya Publ., WB, pp 36–40

Sarjeant WAS (1975) Fossils tracks and impressions of vertebrates. In: Frey RW (eds.) The study of trace fossils. Springer, Switzerland, pp 283–324 (562p)

Simpson S (1975) Classification of tracefossils. In: Frey RW (eds) The study of trace fossils. Springer, Switzerland, pp 39–54 (562p)

Chapter 9
Mangroves Swamp and Tidal-Marsh Sedimentation

Abstract The Thakuran drainage basin is a unique example of a intensely proliferating mangrove ecosystem, and it contains the richest mangrove repository of the Sunderbans. Intertidal zones of newly developed point bars and mid-channel bars besides areas of natural levees and river flood plains are the suitable sites for mangrove flora in the Thakuran Basin. The morphodynamics of the Thakuran tidal river, situated in a tropical meso-macrotidal flow regime, play a vital role in the sedimentation pattern of the tidal marsh and also in changing the mangrove flora of the region. The native fauna of silty marsh sediments often cause changes in tidal currents over the marshes and create a baffling effect against waves and currents. Construction of earthen dams and land reclamation on both banks of the Thakuran River have an adverse effect on the growth of tidal-marsh vegetation.

Keywords Mangrove swamp · Tidal marsh · Species diversity · Tidal sedimentation · Thakuran river · Sunderbans

The Thakuran river channel is sinuous to meandering having connections with many saltwater courses. River point bars, natural levees, river flood plains and mid-channel bars support the creation of mangrove forests and salt marshes. There are three physiographic states for mangrove settings along the Thakuran drainage basin: (i) point bars and mid channel bars as areas of accretion leading to lateral and/or vertical accumulation of land forms, (ii) newer levees and outer concave banks as areas of erosion leading to removal or lowering of surfaces and (iii) older levees as sites of a quasi-equilibrium state of erosion and deposition. Of these the accretionary point bars seem to be the most suitable sites for the occurrence of the mangrove forests. Margins of mature mid-channel bars occupy the next important position. Subareal levees and distal flood plains are rather suitable sites for the marshes. Geomorphic environment helps to specify the community character of mangrove forests, and texture of mangrove substrates appears to be very environment sensitive. Studies of the textural parameters of sediments throw much light on the physical and hydrodynamic conditions of sediment transport, erosion and deposition in mangrove areas.

G.K. Das, *Tidal Sedimentation of the Sunderban's Thakuran Basin*,
DOI 10.1007/978-3-319-44191-7_9

9.1 Geomorphological Features of the Mangroves and Marshes

Geomorphologically, each marginal bar can be divided into two important zones;

(i) riverside mangrove swamps in the lower intertidal zone
(ii) tidal marshes in the upper intertidal flats.

The mangrove swamp zones, in almost all cases, show dense mangrove vegetation in the lower-most intertidal region whereas the upper portions are occupied by dwarf mangroves. The dominating mangrove species is *Avicennia alba* together with a few other species such as *Avicennia marina*, *Avicennia officinalis*, and *Sonneratia* sp.

The marshes are characterised by *Porteresia coarctata*, *Suaeda nudiflora, Suaeda maritime* and *Sesuvium portulacastrum* in areas of depression in the upper intertidal flats. *Porteresia coarctata* occurs in patches and maintains close contact with the *Suaeda* species.

9.2 Mangrove Characteristics

Occurrence of mangrove vegetation is a tropical and sub-tropical phenomenon. Luxuriant vegetation is therefore distributed between the Tropics of Cancer and Capricorn. The Thakuran River is characterised by the tropical meso-macrotidal flow regime within the region of the world's largest mangrove forest of the Sunderbans. Mangrove vegetation of the Thakuran drainage basin is inundated with the tidal water twice daily. Soil salinity is a major factor for the growth of the mangrove vegetation. Stunted growth of mangroves is seen in the higher salinity zone in the Thakuran basin. Mangroves have special adaptive features like air-breathing roots known as pneumatophores, stilt roots (Fig. 9.1) or prop roots, root buttress, and plank roots. These features are not only the characteristic features of true mangroves, but they support the growth and abundant occurrence of the mangroves in various dimensions.

These mangroves generally occupy the intertidal zones of newly developed point bars and mid-channel bars besides areas of natural levees and river flood plains. Mangrove forests are truncated where embankments occurs in the flood plains behind intertidal zones. The lower-most intertidal zone generally is densely forested, whereas the upper portion is occupied by dwarf mangroves. The regular inundation and exposure, together with a freshwater discharge from the upland areas during monsoon, are considered to be the factors for the healthy growth of mangroves in the lower intertidal flats. The Thakuran drainage basin is a unique, highly productive mangrove ecosystem and contains the rich mangrove repository of Sunderbans. The mangroves of the Thakuran drainage basin comprising 64 species under 31 families (Table 9.1) belong to (i) true mangroves species or major element of mangroves,

Fig. 9.1 Stilt roots of mangroves inundated during flood tide

(ii) mangrove associated plants or minor relatives of the mangroves, (iii) back mangrove trees and shrubs, (iv) non-halophytic and non-mangrove associates, (v) halophytic herbs, shrubs and weed flora and (vi) epiphytic or parasitic shrubs and ferns growing on the mangrove trees (Sharma and Naskar 2010).

9.3 Tidal Sedimentation on Mangrove Swamps

The process of erosion and deposition of the meandering tidal rivers is primarily controlled by transverse helicoidal flows coupled with the landward and seaward movement of flood and ebb currents acting along the course of the river. The Thakuran tidal river of the Sunderbans offers an example that helps to understand the principle of sedimentation in such a complex system and its impact on the bordering mangrove vegetation.

The morphodynamics of the Thakuran tidal river, particularly the change in its bank and adjoining landforms, play a vital role in changing the mangrove flora of the Sunderbans of West Bengal, India. Data collected along and across the river reveal four different mechanisms of bank erosion. These are: (i) erosion due to the impact of longitudinal and helicoidal flows in the meandering stream, (ii) the thrust of the ebb and flood cycles and fluctuation of high tidal amplitude, (iii) cohesion of

Table 9.1 Mangroves species of the Thakuran drainage basin

Family	Sl. no	Local name	Scientific name	Type	Halophytic adaptive features
Rhizophoraceae	1.	Garjan	*Rhizophora mucronata* Lamk.	Tree	Prop/stilt root, Viviparous germination
	2.	Bhora	*Rhizophora apiculata* Blume	Tree	Prop/stilt root, Viviparous germination
	3.	Jele Goran	*Ceriops decandra* Griff	Tree	Fused stilt root, Viviparous germination
	4.	Mat Goran	*Ceriops tagal* Perrottet	Tree	Fused stilt root, Viviparous germination
	5.	Bakul	*Bruguiera cylindrica* (L) Blume	Tree	Broom like roots, Viviparous germination
	6.	Kankra	*Bruguiera gymnorhiza* (L) Lamk.	Tree	Broom like roots, Viviparous germination
	7.	Kankra	*Bruguiera sexangula* (Lour) Poirot	Tree	Broom like roots, Viviparous germination
	8.	Champa	*Bruguiera parviflora* Roxb.	Tree	Stilt roots, Viviparous germination
	9.	Garia	*Kandelia candel* L.	Tree	Stilt roots, Viviparous germination
Avicenniaceae	10.	Piara Baen	*Avicennia alba* Blume	Tree	Pneumatophore, Cryptoviviparous
	11.	Kala Baen	*Avicennia marina (F)* Vierh.	Tree	Pneumatophore, Cryptoviviparous
	12.	Jat Baen	*Avicennia officinalis* L.	Tree	Pneumatophore, Cryptoviviparous
Sonneratiaceae	13.	Kaora	*Sonneratia apetala* Buch-Han.	Tree	Pneumatophore, Incipient Viviparous
	14.	Chak-kaora	*Sonneratia caseolaris* L.	Tree	Pneumatophore
	15.	Ora	*Sonneratia alba* Smith	Tree	Pneumatophore
	16.	Ora	*Sonneratia griffithii* Kurz	Tree	Pneumatophore

(continued)

Table 9.1 (continued)

Family	Sl. no	Local name	Scientific name	Type	Halophytic adaptive features
Meliaceae	17.	Dhundul	*Xylocarpus granatum* Koen	Tree	Root buttress
	18.	Pasur	*Xylocarpus mekongensis* Pierre	Tree	Root buttress
	19.	Amur	*Aglaia cucullata* Roxb.	Tree	Pneumatophore
Myrsinaceae	20.	Khalsi	*Aegiceras corniculatum* (L) Blanco	Tree	Incipient Viviparous
Plumbaginaceae	21.	Tara	*Aegialitis rotundifolia* Roxb.	Tree	Viviparous germination
Euphorbiaceae	22.	Genwa	*Excoecaria agallocha* L.	Tree	Gall on trunk
Sterculiaceae	23.	Sundari	*Heritiera fomes Buch-Ham*	Tree	Root buttress, Incipient Viviparous
Malvaceae	24.	Bhola	*Hibiscus tiliaceaous* L.	Shrub	–
	25.	Banvendi	*Hibiscus tortuosus* Roxb.	Shrub	–
	26.	Paras	*Thespesia populanea* S.	Tree	–
Combretaceae	27.	Kripal	*Lumnitzera racemosa* Willd	Tree	Looping aerial roots
Palmae	28.	Golpata	*Nypa fruticans* Wurmb	Tree	Incipient Viviparous
	29.	Hental	*Phoenix paludosa* Roxb.	Shrub	Pneumatophores, Incipient Viviparous
Caesalpiniaceae	30.	Nata	*Caesalpinia bonduc* Roxb.	Shrub	Stilt root
	31.	Singrilata	*Caesalpinia crista* L.	Shrub	Stilt root
Apocynaceae	32.	Dakor	*Cerbera odollan* Gaertn	Tree	–
Verbenaceae	33.	Ban Jui	*Clerodendrum inerme* Gaertn	Shrub	Stilt root
Acanthaceae	34.	Hargoza	*Acanthus ilicifolius* L.	Shrub	Stilt root, Incipient Viviparous
	35.	Lata Hargoza	*Acanthus volubilis* Wall	Climber	-do-
Aizoaceae	36.	Jadu Palang	*Sesuvium portulacastrum* L.	Herb	–

(continued)

Table 9.1 (continued)

Family	Sl. no	Local name	Scientific name	Type	Halophytic adaptive features
Tiliaceae	37.	Lata Sundari	*Brownlowia lanceolata* Roxb.	Shrub	–
Papilionaceae	38.	Singar	*Cynometra ramiflora* L.	Tree	Stilt root
	39.	Kalilata	*Derris scandens* Benth.	Climber	Stilt root
	40.	Karanja	*Derris indica* Bennet	Climber	-do-
	41.	Panlata	*Derris trifoliata* Lour.	Climber	-do-
	42.	Chuliakanta	*Dalbergia spinosa* Roxb.	Climber	-do-
	43.	Gila	*Entada scandens* L.	Climber	–
Asclepiadaceae	44.	Dudhilata	*Finlaysonia obovata* Wall.	Climber	–
Chenopodiaceae	45.	Nona Hatishurh	*Heliotropium curassavieum*	Herb	–
	46.	Giria Sak	*Suaeda nudiflora* Roxb.	Herb	–
	47.	Giria Sak	*Suaeda maritima* Dumort	Herb	–
	48.	Nona Sak	*Salicornia brachiata* Roxb.	Herb	–
Poaceae	49.	Dhani ghas	*Porteresia coarctata* Takeoka	Herb	–
Tamaricaceae	50.	Nona Jhau	*Tamarix gallica* L.	Shrub	Stilt root
	51.	Bon Jhau	*Tamarix troupii* Hole	Shrub	–
	52.	Lal Jhau	*Tamarix dioica* Roxb.	Shrub	Stilt root
Lauraceae	53.	Manda	*Viscum orientale* Willd.	Climber	–
Convolvulaceae	54.	Chhagal Kunri	*Ipomoea pescaprae Sweet*	Herb	–
Rubiaceae	55.	Tagri bani	*Scyphiphora hydrophyllacea* G.	Shrub	Wavy aerial root
Rutaceae	56.	Banlebu	*Atalantia correa* M.	Shrub	–

(continued)

Table 9.1 (continued)

Family	Sl. no	Local name	Scientific name	Type	Halophytic adaptive features
Asclepiadaceae	57.	Bowle Lata	*Sarcolobus globosus* W.	Shrub	–
	58.	Bowle Lata	*Sarcolobus carinatus* Wall.	Shrub	–
	59.	Manda Lata	*Hoya parasitica* Wall.	Shrub	–
Araceae	60.	Kerali	*Cryptocoryne ciliata* Roxb.	Herb	–
Amaryllidaceae	61.	Sukh Darshan	*Crinum defixum* K. G.	Herb	–
Ruppiaceae	62.	Nona Jhajhi	*Ruppia maritima* L.	Herb	–
Loranthaceae	63.	Pargachha	*Dendrophthoe falcata* L.	Shurb	–
Pteridaceae	64.	Hodo	*Acrostichum aureum* L.	Shrub	Fern, rhizomes

grains attributed by their textural parameters, e.g. banks composed of sand are eroded more than these of clay or more cohesive material and (iv) different degrees of subsidence causing piecemeal bank erosion and subsequent modification by waves near the mouth of the river.

The low alluvial plain sediments in this stretch of the delta are reworked and recycled to generate the various morphological units and subunits of landforms all along the river basin. The neap-spring cycles, the flood-ebb cycles and the tidal regime of the river have influenced the configuration of the marginal levees, accretionary point bars, mid-channel bars and river mouth bars. The process of accretion in the Thakuran River is controlled mostly by a higher rate of sediment trapping in areas of geomorphic highs with networks of mangrove vegetation. Subsidence of the bank with simultaneous sediment filling all along the course is quite common. Evidences of subsidence are marked by: (i) erosion of riverbanks, (ii) undercutting of mangroves, (iii) exposures of chunks of mangroves and (iv) supratidal bluff breccias in the intertidal stratigraphy. The banks of the Thakuran register evidence of sagging in certain portions, alternating with gentler sloping areas to the order of 3°–5°.

9.4 Tidal Marshes

Marsh vegetation occurs sporadically all along the river margins, particularly in the muddy upper intertidal zone to the supratidal zone. They generally occupy areas above mangrove swamps and are typical of intertidal zone behind embankments or earthen dams constructed for protecting inland areas from salt-water menace. In addition to their diverse ecological and economical importance, marsh vegetation has a special role as sediment baffles (Davis 1983; Das 2015). However, embankments on both banks, above the extension of marginal bars of the river and particularly in the reclaimed area, stunt the growth of tidal marsh vegetations in the Thakuran drainage basin.

The linear intertidal zones, ranging in width between 250 m and 300 m and bordering the distributary channel systems of the tidal Thakuran River, are typically low wetlands covered with marsh grasses like *Porteresia coarctata* and halophytic herbs like *Suaeda nudiflora, Suaeda maritima* and *Sesuvium portulacastrum.* The vegetation carpet along with a variety of benthic organisms, viz. *Telescopium telescopium*, *Anadara granosa*, crabs like *Uca acuta* and *Ocypoda macrocera*, *Nerita* sp., *Cerithedea* sp. and hermit crabs, modify wave and current processes. This results in decreased energy in or near the muddy substrate ($M_Z = 5.41$–7.01 phi) creating a baffling effect. The marshes in all cases are fronted by dwarf and tall mangrove swamps located between low and high neap-tide levels (Fig. 9.2). The efficient baffling and sediment trapping mechanism imparts a greater coherence since the precipitated particles are prevented from being removed. Apparently, no bedforms are recognisable except through surface bioturbation in the form of traces, trails and borings of animals. Parallel laminations are the only internal structures often affected by press marks and burrows of animals and the root systems of marsh vegetation.

Fig. 9.2 Occurrences of the mangroves and tidal marsh (*Porteresia coarctata*) complex on the marginal bar along the stretch of the river

9.5 Intertidal Marsh Sedimentation

Tidal marshes are important both in understanding the hydrodynamics of sedimentation and the role of vegetation as sediment baffles. Very few studies have been undertaken on the sedimentation pattern and morphological character of marshes from the coastal plains in India. However, excellent summaries of marsh sediments and sedimentation history have been given by Edwards and Frey (1977), Frey and Basan (1978), Davis (1983) and Orson et al. (1990). By definition, marshes are coastal wetlands or river-margin wetlands, and they have recently been much emphasized by ecologists in order to exploit such areas for multiple purposes.

9.6 Texture of the Marsh Sediments

Texturally the marsh sediments are composed of 90 % silt, with equal proportions of fine sand and clay forming the remaining 10 % (Table 9.2). The marsh sediments are mostly moderately well sorted to moderately sorted ($\sigma_1 = 0.57–1.4$) with a graphic mean size belonging to the silt fraction ($M_Z = 5.41–7.01$ phi). Most samples show a slight tendency to negative skewness. The negative skewness in the present situation perhaps indicates trapping of larger bed-load particles by the marsh vegetation. The cumulative curves show mostly steep peaks with K_G values of greater than 1.5 (Table 9.3). The statistical size parameters of the sediment samples are very much analogous to the lagoon or distal-shelf sediments (Friedman and Sandars 1978). Inflections in the cumulative curves at 5.0–6.0 phi is a statistical fact for all the samples. All the cumulative curves are non-linear with zig-zag patterns and bear close resemblance in pattern with one other. The salt-marsh sediments under conditions of exposure during the neap tide often register salt encrustation on the surface. The pH range of salt-marsh sediments is 6.73–7.69 and the salinity range is 2.86–5.36 ppt (Table 9.4).

Table 9.2 Proportions of sand, silt and clay in the marsh sediments

Sample no.	Sand %	Silt %	Clay %
K_4	0.80	98.80	0.40
K_5	1.32	94.48	4.20
K_6	6.89	90.32	2.79
K_7	3.44	95.44	1.12
K_8	0.16	96.12	3.72
K_{11}	18.12	78.12	3.72
K_{12}	2.36	97.53	0.11
K_{13}	1.36	94.88	3.76

Table 9.3 Statistical size parameters of the marsh sediments

tt	Inclusive graphic mean size (M_Z)	Inclusive graphic standard deviation (σ_1)	Inclusive graphic skewness (SK_1)	Inclusive graphic kurtosis (K_G)
K_4	6.05	0.84	−0.10	1.37
K_5	6.31	0.87	0.45	2.17
K_6	6.76	1.40	−0.37	2.08
K_7	6.50	0.59	−0.18	1.93
K_8	5.41	0.78	0.35	3.17
K_{11}	5.46	1.19	−0.69	0.97
K_{12}	7.76	1.35	−0.38	0.78
K_{13}	7.01	0.61	−0.41	1.53

Table 9.4 Salinity and pH of the marsh sediments

Sample no.	pH	Salinity in ppt (‰)
K_4	7.30	4.09
K_5	7.68	5.36
K_6	7.27	5.04
K_7	6.75	3.44
K_8	7.05	3.44
K_{11}	7.69	2.86
K_{12}	6.73	3.44
K_{13}	6.75	3.16

9.7 Physical and Biogenic Structures

In order to examine the marshy field in three dimensions, some L-shaped trenches were cut. Excepting cases of obscured parallel laminations (Harms and Fahenstock 1964), no other structure was recognisable. These laminations were further affected variously by the root systems of the creepers and grasses and by benthic-animal burrows. The marsh flat is often inhabited by a variety of benthic organisms like gastropods, *Telescopium telescopium*, *Nerita* sp., *Cerithedea* sp., bivalves like *Anadara granosa*, crabs like *Uca acuta* and *Ocypoda macrocera*, hermit crabs and *Sipunculas* species. These benthic forms yield various bioturbation features like tracks, trails, press marks and borings of variable sizes and length (Reineck and Singh 1980). Surface physical structures are rare or absent.

The tidal marshes along with the inhabiting fauna cause some modifications to the flashing paths of the tidal currents over the marshes. This results in a decreased energy condition on the surface of the muddy substrates and creates a baffling effect against the waves and currents (Davis 1983). The marshes are flooded twice daily during the high-water spring to the high-water neap tidal cycles with sediment-laden water. During this span of time, the fine suspended particles of very fine sand, silt and clay are trapped onto the marsh surface and settle down gradually (Lundgren 1986). The marshes, with their occurrence in the marginal bars, are fronted by mangroves which impart a greater coherence for the precipitated particles. Thus, under condition of intense waves and current during storms, the fine particles of the marshes are prevented from being removed. It is hypothesized that the sediment accumulation in the marshes would encroach on the mangrove swamps with time, and these sediment accumulations with tidal cycles would change its character and estuarine configuration.

9.8 Summary

The bank morphology with terrigenous-sediment settings along the Thakuran River is modified by subsidence and sediment compaction which have contributed to a terraced pattern supporting mangroves at different stratigraphic levels. Mangroves of different generations occupy different sub-environment realms like mid-channel bars, point bars, natural levees and channel flood plains in a semi-diurnal, meso-macrotidal setting with strong bidirectional currents. These various geomorphic zones act as areas supporting mangrove forests and salt marshes. Of these, the accretionary point bars seem to be the most suitable sites for the occurrence of mangrove flora. Margins of mature mid-channel bars occupy the next most important position. Several levees and distal flood plains are rather more suitable for the tidal marshes.

References

Das GK (2015) Estuarine morphodynamics of the Sunderbans. Coastal research library, vol 11. Springer, Switzerland, 211p

Davis RA Jr (1983) Depositional system: a geomorphic appearance to sedimentary geology. Prentice Hall, 669p

Edwards JM, Frey RW (1977) Substrate characteristics within a Holocene salt marsh of Sapelo island. Georgia Senckenbergiana Merit 9:215–259

Frey RW, Basan PB (1978) Coastal sedimentary environments. Springer, New York, pp 101–169

Friedman GM, Sandars JE (1978) Principles of sedimentology. Wiley, New York, p 792p

Harms JC, Fahenstock RK (1964) Stratifications, bedforms and flow phenomena. Bull Am Asso Petrol Geologist 48:530–544

Lundgren L (1986) Environmental geology. Prentice Hall, 576p

Orson RA, Simpson RL, Good R (1990) Rates of sediment accumulation in a tidal fresh water marsh. J Sed Petrol 60:859–869

Reineck HE, Singh IB (1980) Depositional sedimentary environments. Springer, New York 408p

Sharma AP, Naskar KR (2010) Coastal zone vegetation in India with reference to mangroves and need for their conservation. In: Naskar KR (ed) Sundarbans issues & threats. Central Inland Fisheries Research Institute (ICAR). Spl Publ, p 9–40 (184p)